Ideology, Science and
Human Geography

Ideology, Science and Human Geography

Derek Gregory

Fellow of Sidney Sussex College
and Lecturer in Geography,
University of Cambridge

St. Martin's Press
New York

ISBN 0-312-40477-8

Library of Congress Cataloging in Publication Data

Gregory, Derek, 1951–
    Ideology, science, and human geography.

    Bibliography: p.
    Includes index.
    1. Geography–Philosophy.   I. Title.
G70.G73 1979   910′.01   78–11746
ISBN 0-312-40477-8

For my parents

Contents

Figures

Preface

In this book I have tried to develop an alternative conception of science on which our inquiries might be based, which involves – in very general terms – a transition from a traditional or a positive conception to an explicitly critical one. The enterprise is scarcely novel: Bartels has already characterized the history of geography in terms of successive 'waves of rationality', and has drawn on Habermas to identify a fourth, critical surge currently convulsing the subject; Claval has turned to Foucault, to suggest a progression of descriptive, prescriptive and ultimately critical *épistémés* moving through the subject; Harvey has argued in classical Marxian terms for a more clearly defined critical reorientation within the subject; and Olsson has advocated, through a revitalized linguistic philosophy, the construction of an avowedly critical geography. All of these are preliminary recommendations, of course, and each has its own difficulties, but although they do not speak with one voice they all draw geography out of its ignoble (and in any case ineffective) isolation and plunge it back into the vital conduct of practical life.

The discussions which follow are concerned for the most part with the conditions which might make this jump possible: that is to say, they are *epistemological*, and to pose questions at this abstract level presents major difficulties for a critical science which must find its ultimate legitimation in practical life. Even so, I am disposed to agree with Anuchin that it is the concentration on theory that safeguards the practical importance of science and, as a result, that it is strategically necessary to clarify the foundations of our theoretical activity.

But it is clearly not enough to remain at this abstract level: life has to be breathed into these abstract concepts in order, as Marx once put it, to 'force the frozen circumstances to dance by singing to them their own melody'. There is a sense in which the orchestra is, as it were, still sorting out the theoretical score, but beneath its less and less fragmented chords resonate the rhythms of the actions to which the theories must be tied.

In the past, I suggest, geography circumscribed these actions by drawing back from an interrogation of its *problematic*, its system of concepts: categories were used with their connotations still-born, silent within the texts. It was not so much that practical consequences never flowed from the corpus of meaning to which geographers subscribed at any one time – quite the reverse – but rather that there was a conspicuous failure to specify what these were and to identify the mediations between theory and practice. And, in large measure, this was because, as Pred has indicated, geographers were less concerned with problems of intellectual substance than with the possible 'geographical' ramifications of models, concepts and techniques used by other scholars, a predilection which effectively prevented them from the 'non-geographical' evaluation of their various borrowings.

In reclaiming geography from the academic pawnbroker, however, I do not intend to follow Pred in calling for the development and deployment of our *own* conceptual structures: I cannot imagine what these would consist of. Instead, I prefer to connect geography to the social sciences in general and to attempt an examination of discourse which must necessarily transcend the artificial and historically specific confines of any one discipline. *Ideology, Science and Human Geography* is addressed to a particular audience, therefore; it does not present a particular play.

Derek Gregory
Sidney Sussex College, Cambridge
1 September 1977

Acknowledgements

It has proved difficult to stop writing this book. The more I discussed the ideas and issues which it was supposed to address, the more additions and excisions crowded in. My first debt, therefore, is really a joint one: to the students whose counter-arguments, questions and criticisms have contributed in no small way to what I have started to learn about the social sciences, and to Rob Shreeve for somehow putting up with the missed delivery dates and the manuscript alterations which resulted from this never-ending exploration.

I am also grateful to my colleagues at Cambridge for their interest and advice, and for their willingness to respond so openly and in such detail to notions often at odds with their own. Some of the arguments contained in the following pages were first presented in a series of Occasional Discussions in Historical Geography, where I was forced to justify many of my assumptions and to clarify most of my formulations – something I still haven't finished – and the existence of such a painfully honest forum has proved to be a constant and rewarding stimulus; one which also ensured that I didn't stray too far from the practical business of research.

I owe a particular debt to Mark Billinge, and I am only too conscious of how inadequate an acknowledgement this is of a friend whose scholarship and encouragement did much to keep everything in perspective.

Michael Chisholm, Tony Giddens and Chris Hamnett kindly read an early version of the complete manuscript, and suggested many revisions which I tried to incorporate wherever possible. More particularly, Alan Baker, Richard Smith, David Stoddart, Nigel Thrift and Tony Wrigley helped me to sort out my conception of structural explanation; Mark Billinge, Anne Buttimer, Gunnar Olsson and Hugh Prince commented freely on my attempts to understand something of the nature of reflexive explanation; and Martin Boddy, Peter Cromar, Jim Lewis and Tim Unwin guided me towards a deeper recognition of what committed explanation ought to be about. I suspect I have only imperfectly grasped what they were arguing, but I am grateful to all of them for their help.

Pamela Lucas created order out of the chaos of my hastily drawn thumbnail sketches, frequently at short notice but always with patience, skill and good humour, and Jean Reddy, Peg Smith and the indefatigable Mrs Powell, in their different ways, made me realize how much I owed to good libraries.

I also want to thank Mark, Natasha, Roger, Alison, Mike, Jo, Steve and Anne: they will know why. And, finally, I owe more to Liz than to anyone – and far more than I have ever found the words to say.

D.G.

Introduction:
A place among the natural sciences

When Charles had quenched his thirst and cooled his brow with his wetted handkerchief he began to look seriously around him. Or at least he tried to look seriously around him; but the little slope on which he found himself, the prospect before him, the sounds, the scents, the unalloyed wildness of growth and burgeoning fertility, forced him into anti-science. The ground about him was studded gold and pale yellow with celandines and primroses and banked by the bridal white of densely blossoming sloe; where jubilantly green-tipped elders shaded the mossy banks of the little brook he had drunk from were clusters of moschatel and woodsorrel, most delicate of English spring flowers. Higher up the slope he saw the white heads of anemones, and beyond them deep green drifts of bluebell leaves. A distant woodpecker drummed in the branches of some high tree, and bullfinches whistled quietly over his head; newly arrived chiffchaffs and willow-warblers sang in every bush and treetop. When he turned he saw the blue sea, now washing far below; and the whole extent of Lyme Bay reaching round, diminishing cliffs that dropped into the endless yellow sabre of the Chesil Bank, whose remote tip touched that strange English Gibraltar, Portland Bill, a thin grey shadow wedged between azures. . . .

Science eventually regained its hegemony, and he began to search among the beds of flint along the course of the stream for his tests. He found a pretty fragment of fossil scallop, but the sea-urchins eluded him. Gradually he moved through the trees to the west, bending, carefully quartering the ground with his eyes, moving on a few paces, then repeating the same procedure. Now and then he would turn over a likely-looking flint with the end of his ashplant. But he had no luck.

John Fowles: The French Lieutenant's Woman

The Victorians accorded a special privilege to the canons of natural science, and the first steps of academic geography in Britain were made in the boots of a severely natural science tradition. Its early practitioners eschewed any close involvement with man himself, and although scattered and precocious references to the work of pioneering social scientists like Comte and Spencer can be found in the writings of some Victorian geographers their efforts never diverted the mainstream of geographic thought. The party line was captured most precisely by Sir Halford Mackinder. For him the basis of geography was the physical environment, and when he admitted that 'the other element is, of course, man in society' this was relegated to a footnote in which he observed that 'the analysis of this will be shorter than that of the environment' (Mackinder 1887, 143). Less than twelve months later the President of the Royal Geographical Society claimed for the discipline 'a place among the natural sciences' (Strachey 1888, 151). He emphasized the methodological unity of all science:

its methods, though first developed by the study of mathematics and of the physical forces of nature, are applicable to all the objects of our senses and the subjects of our thoughts. The foundation of all knowledge is the direct observation of facts; by applying thought to the facts thus observed, we seek through a process of classification and comparison for the causes of which the observed phenomena are the results, and the conclusions thus obtained constitute science [Strachey 1888, 149].

Assertions like this were by now a commonplace, however, and Strachey's attempts to develop the thesis made little impact on his Cambridge audience (Stoddart 1975a, 223). What was remarkable was that the commitment to methodological unity produced not a dialogue between physical and human geography, but a monologue to which the latter responded submissively. It was of course true that the social sciences themselves were cast in the shadow of the natural sciences, but this was in the methodological sense of securing an identical *logical* structure. Even when human geography assumed a more coherent form the two sides of the coin were not of the same currency. Ten years after Mackinder's lecture Scott Keltie could be found assuring the members of the Royal Geographical Society that 'man is the ultimate term in the geographical problem' (Keltie 1897, 310), but although this claim now appeared in the body of the address he joined his predecessors in making it clear that the dimensions of the problem and the insertion of man into it were to be governed by the precepts of an essentially physical geography. Most geographers still followed Mackinder in believing that 'no rational [human] geography can exist which is not built upon and subsequent to physical geography' (Mackinder 1887, 143).

If man had to be embraced, then the scene was going to be played with both feet firmly on the ground throughout.

This early reluctance to consider man in society was prompted at least in part by an important search for intellectual respectability, and a belief that by structuring human geography in terms of physical geography such a goal could be attained. Given the difficulties of persuading the universities that geography was a distinctive discipline worthy of serious study in its own right, so that it should be included in their teaching programmes, its protagonists were obliged to demonstrate an almost messianic commitment to its proper conduct and its widest promulgation. At the inauguration of the Southampton Geographical Society, for example, Sir Clements Markham spoke of the need to attract young men to 'impart their enthusiasm far and near'. He thought of them 'mounting their bicycles, with lanterns and slides strapped on their backs, [to] rush up the valleys of the Itchen, the Test, the Avon, the Hamble, and penetrate to the remotest villages of the Meonwara'. In this way, he declared, 'South Hampshire would become a geographically enlightened region, with your society as its centre' (Markham 1898, 13). And from centres like Southampton the new 'scientific' geography would spread, commanding both popular and scholarly support as it came to be identified with the natural science doctrines which had already fired the imagination of late Victorian society.

But there was another strand to this argument, because if science was one of the brightest stars in the Victorian sky, utility was its twin. Douglas Freshfield had already castigated those 'who still say that geography does not "pay" ' by claiming that 'for what is justly held in private schools as an end and object of education, getting a boy on the foundation of Eton, geography may prove more useful than they think'. He said this having in his pocket 'one of the last examination papers, in which there are several excellent geographical questions' (Freshfield 1886, 704). This appeal to Eton carried with it an obvious recognition of the Empire which radiated from its playing-fields, and Freshfield demanded whether 'we English who inherit so large a part of the world [shall] not acquaint ourselves with our inheritance and the conditions under which we can retain and make the most of it?' (Freshfield 1886, 701). And here again it was the contribution of a specifically 'scientific' geography which was invoked. The affirmation of the ideology of industrial capitalism which underpinned this was most clearly revealed by Sir

George Robertson's description of it as 'the Science of Distances – the science of the merchant, the statesman, and the strategist' (Robertson 1900, 457), a characterization which appeared to make a command of geography vital both for the maintenance of the Empire itself and for the ascent of men to the most acclaimed positions of profit and power within it.

These early developments, even in the crude form in which they are presented here, are important to the argument which I want to explore in the following chapters for two reasons. In the first place it is important to realize how much geography as we know it today is the product of a series of decisions, some considered and others more impulsive, some prompted by idealism and others more pragmatic, but all of them taken in particular historical situations by a relatively small number of men occupying positions of authority and prestige within the discipline. Geography was not so much what geographers did as what they were obliged to do, and while their contributions are manifestly worthy of careful examination they should not lead us into attributing some kind of inalienable rationale to the succession of changes which the discipline underwent as a result of their actions. It is only too easy to approach their various efforts through an intellectual history which emphasizes an abiding consensus within geography itself, and then establishes some points of contact between its mode of thought and those held to obtain in other judiciously selected and similarly united disciplines. Such exercises are obviously reassuring, because they often simply confirm the viability of the *status quo*, joining our efforts with those of others to erect a common orthodoxy from which there can be no legitimate dissent. As Alvin Gouldner warned, 'the search for convergences with and in the past seeks to reveal a tacit consensus of great minds and, by showing this, to lend credence to the conclusions that they are held to have converged upon unwittingly. Convergence thus becomes a rhetoric, a way of persuading men to accept certain views' (Gouldner 1971, 17).

Any points of contact are naturally very important, but if we are to evaluate them and achieve any real awareness of our own position, then we also need to consider points of conflict and the alternative views which they commend to us. We need to examine not only differences within geography itself, but also to investigate those paths which cut across or diverged from our own somewhere along the way to inculcate flourishing bodies of thought which direct their attentions towards problems similar to our own, but from very different perspectives. In other words I am suggesting that we regard our own preconceptions and procedures as but one set among a much larger (and in some ways a much richer) universe, and so consider very closely our reasons for acting in the way we do. I am not advocating the kind of tolerance which might be interpreted as eclecticism or kleptomania, depending on your point of view: geography has seen too many ventures of that kind. I am saying that our adoption of one position rather than another is much more convincing if the choice is a conscious one, deriving from a careful appraisal of the alternatives. In short, 'we have to admit a distinction between what should be geographical in kind and what is presently geographical in format. The latter, representing the conventions of the profession, should not be permitted to define the former' (Hägerstrand 1973, 67).

In the second place, and in a sense revealing the way in which a continuity of thought was indeed internalized and legitimated by (or at least on behalf of) the discipline, it is important to reflect on how truly Victorian were the men who pioneered the New Geography of the 1950s and 1960s – the 'men of action [who] continued to play with the old prizes' (Olsson 1975, 476). If their words sound hollow today it is not because, in Brookfield's phrase, they are Young Turks become Old Mamelukes, but because they have always been responding to the cadences of those early generations, whether they were conscious of it or not. To the extent that this is true, the periodic performance of anthems to the natural sciences assumes the force and form of myth itself, which 'appears as conductor of an orchestra of which the listeners are the silent performers' (Lévi-Strauss 1964, 25). A first task is to uncover the structure of this myth.

For the moment I simply want to assert that mid-twentieth-century geography formalized the ideals of its Victorian forebears. I am prepared to accept that it may even have extended them, in so far as there were several strands to the philosophical thread spun across the intervening decades, but I am certainly not prepared to accept that it transformed them. By reaffirming its commitment to the natural sciences, geography embraced a philosophy which Strachey would have had little difficulty in recognizing. What was distinctive about it was that the earlier emphasis on the ontological primacy of the natural world was replaced by the epistemological primacy of the natural sciences. This meant a tacit allegiance to a positivist philosophy of science, which in its original Comtean form grounded all knowledge in the direct experience of an immediate reality: from this could be derived the laws regulating both man and his material universe. The concern with the discovery of general laws, or at least with the formulation and verification of particular theories (Wilson 1972, 32), was an important one because it indicated that the methods by which other sciences had secured intellectual recognition might work for geography as well. What was more, those methods appeared to provide an essential technical foundation for the elaboration of private and public policy. Geography could now prescribe 'the *optimum means* of achieving a *given set* of social objectives' (Berry 1972, 80), so that more research funds could be attracted from corporations and the state itself. In short, the Victorian Pantheon was refurbished.

To be sure, the estate agent's description was soon qualified by the tenants, and there were many who would not even cross the threshold in case they fell through the floorboards. Grants were secured to make necessary internal improvements, all of which were advertised in a literature which was promotional in more ways than one, but it was not long before the first tenants moved out. Berry, who had been one of the first to move in, now spoke of his sense of frustration, and the need 'to come to terms with the many sources of confusion and doubt about the continuing viability of earlier research orientations' (Berry 1973, 3–4). Hägerstrand declared himself disappointed because geography was 'too incomplete to be able to catch the conditions which circumscribe man's actions' (Hägerstrand 1973, 67), while Harvey, whose *Explanation in Geography* had extended the original lease, recognized 'a clear disparity between the sophisticated theoretical and methodological framework we are using and our ability to say anything really meaningful about events as they unfold around us' (Harvey 1973, 128). In effect, the Pantheon now appeared as a characteristically Victorian folly.

This reappraisal grew out of an increased confidence within the discipline, which encouraged its spokesmen to ask why its (or rather, their) voice was not being heard in the corridors of power. The President of the Institute of British Geographers complained that 'geographers are not being asked to make the contribution of which they are capable', a serious situation since 'universities are now largely funded from the public purse and are subject to an increasing degree of direction by their paymasters'. If their contribution continued to go unrecognized, let alone asked for, would those paymasters disappear, taking their funds with them? (Coppock 1974,

4–5). An editorial in *Area*, the house-journal, had already made it clear that the benefits to be derived from 'more direct participation in the channels of decision-making' came 'not so much from a Snow-like lust for power, as from the general esteem – and the finance – accorded to consultative exercises' (Robson 1972, 213). The implication was obvious. Much the same writing was appearing on North American walls, so that the collective dissatisfaction might in part be attributed to the fear of impotence which puberty brought with it. But it went deeper than this, in so far as an exclusive concern with the survival of the discipline required a degree of pragmatism which many regarded as an abandonment of principle (Smith 1973, 3). The difficulty was that pragmatism was uncomfortably close to opportunism (Harvey 1973, 325). This divorced ends from means, so that the objectives for research were established outside the profession by those paying for it: as such, they were excluded from critical analysis, and it was an abuse of the geographer's privileged professional status to call them into question.

Once this narrowly technical role was rejected and ends were regarded as inextricably linked to means, however, this privileged status was itself called into question. It had its roots in the Victorian myth of the supremacy of the natural sciences, and it had two ritualized expressions: the scientist as observer, and the scientist as hero. The first of these is obvious in its assumption that the scientist is external to the reality he studies, so that neither his own preconceptions nor those of his 'objects of study' are admissible in any legitimate inquiry. The second depends on the first, because if the scientist *can* place himself above society then he alone occupies the vantage point from which to map out a trajectory which those below cannot legitimately challenge (Zelinsky 1975, 141). I regard both of these as erroneous beliefs which depend on the legitimating force of a particular conception of science. In what follows I wish to challenge its hegemony by exploring other modes of social understanding, in the hope that 'we might just develop a sense of wonder, of awe, at the amazing richness of human experience' (Buttimer 1974, 24) instead of collapsing it into abstract and mystifying propositions which offer no prospect of a critical understanding of the relations between man and his material universe.

The task which awaits us is not so very different from the dilemma which confronted Charles on that early March afternoon in 1867.

Part One
Positivism in Geography: Elements of a Critique

1 The positivist legacy and geography

Meeting a friend in a corridor, Wittgenstein said: 'Tell me, why do people always say it was *natural* for men to assume that the sun went round the earth, rather than that the earth was rotating?' His friend said, 'Well, obviously, because it just *looks* as if the sun is going round the earth.' To which the philosopher replied, 'Well, what would it have looked like if it had looked as if the earth was rotating?'

Tom Stoppard: *Jumpers*

Less than ten years before Charles sat down to gaze out over Lyme Bay, the Society of Jesus had received a letter from an old and impoverished Frenchman, still mourning the death of his young mistress. The Society was no doubt accustomed to comforting those seeking solace in its strictures, but this letter was very different: signed by '*Le Fondateur de la Religion Universelle, Grand Prêtre de l'Humanité*', it invited the Jesuits to embrace the new 'positive' religion and to work towards the foundation of an occidental positivist republic. Its author was Auguste Comte. His proselytizing zeal had been heightened by the death of his beloved Clothilde, whom he regarded as 'the spiritual symbol of the Virgin-Mother' and whose grave was to be 'a place of sacred pilgrimage' (Manuel 1962, 268), but it could be discerned in a less manic form in his reaction to the Paris barricades of 1848, when he drew up a positive plan of action for the provisional government, and, most important for our present purposes, in his even earlier formulation of positive philosophy itself.

This was first unveiled in a series of private lectures in Paris, given in his apartments on the rue Monsieur-le-Prince, attended by – among others – Alexander von Humboldt, then nearly sixty. Comte's deliberations were interrupted by a severe mental illness, which his physician feared incurable, but by 1829 he was able to resume, and soon afterwards he set about formalizing his *Cours de philosophie positive*. It embraced five methodological precepts: *le réel, la certitude, le précis, l'utile* and *le relative* (Habermas 1972, 74–7).

Le réel meant that the scientific status of knowledge had to be guaranteed by the direct experience of an immediate reality, and this required a particular conception of causality, owing much to Hume, in which causal relations amounted to regular associations between phenomena. Comte claimed that a causal relation 'discovered between any two phenomena enables us both to explain them and foresee them, *each by means of the other*' (Keat and Urry 1975, 73; italics added). The limitation of the terms of a scientific explanation to the phenomenal level therefore meant that the positivist could not have recourse to any supernatural or abstract forces which were by definition outside his direct experience. Indeed, Comte claimed to have regressed through three fundamental stages of knowledge during his breakdown: from his original 'positive' state in which he ordered the world in terms of regular connections between empirically observable phenomena, he reverted to a 'metaphysical' state in which the world was explained in terms of abstract forces, and finally to a 'theological' state in which the world was explained in terms of supernatural forces. During his recovery he came back through this sequence, and he argued that all disciplines were obliged to follow this same succession, from a primitive theological stage to an ultimate positive one. What was more, they would be transformed in a determinate order, working progressively up through a hierarchy of sciences, arranged so that each one addressed an ever more complex level of reality, culminating in the study of society itself. Taking these together, it is easy to see how the study of society, once it became positive, could assume the significance of a universal – if secular – religion: it offered the ultimate understanding of the highest order of reality known to man. To be sure, it was a bounded understanding, and Comte did not believe that *all* questions could be answered in positivist terms. What he did claim was that 'if they could not be so answered then they could not be answered at all – so there was no point in asking them' (Bryant 1975, 401). In effect, then, his insistence on *le réel* effectively proscribed any critical reflection on the principles of science.

This rule of phenomenalism (Kolakowski 1972, 11) had to be complemented by *la certitude*, which meant that the scientific status of knowledge had to be guaranteed by the common experience of reality, a mode of apprehension which was accessible to all scientists and which ensured the replicability of their observations – in other words, the unity of the scientific method. It followed from this that disciplines were to be distinguished by their object of study, and not by their method. This in turn required *le précis*, which defined the scientific method as the formal construction of theories whose consequences could be tested in some way. Value-judgements were immediately excluded from scientific inquiry because they were ethical assertions rather than empirical predictions, incapable of verification. *L'utile* confirmed this by regarding all scientific knowledge as technically utilizable, concerned with 'means' not 'ends', and although in one sense this could be interpreted as nothing more than instrumentalism its main function in the Comtean schema was to rule value-judgements out of the scientific court (Keat and Urry 1975, 73).

This should perhaps be qualified, though, because Comte conjoined *le précis* and *l'utile* in a distinctive way. His scientific method was intended to reveal the laws of coexistence and succession which governed society, and he maintained that these allowed no variation; once men realized this, then the battalions and barricades would be removed from the map of Europe. As he put it himself, 'there is no chance of order and agreement but in subjecting social phenomena, like all others, to invariable laws, which shall as a whole prescribe for each period, with entire certainty, the limits and character of political action' (Comte, in Bryant 1975, 400). If the scientist was obliged to accept society as these laws dictated it to be, however, *he did not have to accept society as it was*, particularly when he looked out on a continent disrupted and corrupted by politicians who evidently did *not* recognize the inexorable force of these social laws (Gouldner 1971, 101). While certainly excluding value-judgements which were not in accordance with these laws, therefore, Comte's positivism was also distinguished by what Frisby (1976, xii) calls an affirmative impulse of critical enlightenment, in the sense of seeking to remove the misery which resulted from misguided attempts to force society down paths which were necessarily closed to it. This impulse was regarded as non-ideological in so far as the positivist was not required to assume any ethical position in order to demonstrate the truth of his statements or the invincibility of the social laws which they embodied (Fay 1975, 65).

Finally, *le relative* depicted scientific knowledge as unfinished and relative, progressing by the gradual unification of theories which would consolidate man's awareness of the social laws. It followed from this that scientific progress was paralleled by social progress, and Comte tied his three stages of knowledge to three 'modes of action': the theological to the military, the metaphysical to the defensive, and the positive to the industrial.

From this fundamental correlation there also results the general explanation of the three natural ages of humanity. Its long infancy, which fills the whole of antiquity, had to be essentially theological and military; its adolescence in the Middle Ages was metaphysical and feudal; finally, its maturity, which has only begun to be appreciated in the last few centuries, is necessarily positive and industrial [Comte, in Manuel 1962, 279].

Comte was neither the first nor the last to devise a tripartite history of civilization, of course, but his proposals show that he saw positivism very much as a child of the nineteenth century. It is important not to lose sight of the many changes (and challenges) which it has seen since then. I do not wish to trace all of these through, but I do want to consider how positivism's formal translation into geography related to these original precepts and, in particular, whether its clarification or emendation of them can be accommodated within other more recent positivist traditions.

The discussion which follows takes each of Comte's precepts in turn and presents some of their geographical counterparts. The equivalences are not always exact, and there is the ever-present risk of selective quotation, of arguing a case by drawing together heterogeneous statements made by geographers who never accepted the assumptions made on either side of them but which are nevertheless ascribed to them by simple association. As Lewis and Melville (1977, in press) note, 'sooner or later you can discover the "right" quotation in the literature and, even if the viewpoint is not shared by the majority of the practitioners, it may serve its purpose in creating the desired image'. Their solution, that of fastening on just one set of pronouncements, is obviously not open to us; we must proceed with caution, therefore, and this means making careful qualifications wherever necessary, even at the risk of softening the outlines of the argument as a whole.

Positivist explanation in geography

In his classic essay on 'The Morphology of Landscape', published in 1925, Carl Sauer represented geography as 'a science that finds its entire field in the landscape'. According to him 'the systematic organisation of the content of the landscape proceeds with the repression of *a priori* theories concerning it', so that geography relies on 'a purely evidential system, without prepossession regarding the meaning of its evidence'. It is concerned only to establish 'the connections of the phenomena' in the visible landscape, and these connections are ones of spatial association and not, definitely not, of some hidden causality. 'Causal geography', he declared, had had its day and geography could now be established as a 'positive science'. He certainly understood this in a Comtean sense, inasmuch as he upheld Goethe's claim that 'one need not seek for something beyond the phenomena; they themselves are the lore (*Lehre*) [laws]'. In effect, therefore, his 'morphologic method' was simply a restatement of Comte's designation of *le réel* as the only appropriate domain for scientific inquiry.

Against this interpretation Relph (1970, 195) maintains that Sauer's attempt to formalize 'a critical system which embraces the phenomenology of landscape, in order to grasp all its meanings' prefigures the phenomenology of the Husserlian school, which would deny the possibility of an immediately universal experience of 'objective' reality and which would instead be articulated around introspective images of a 'subjective' world (see below, pp. 127–9). This, I think, misses Sauer's point. He certainly accepted that 'a good deal of the meaning of area lies beyond scientific regimentation' and, as his famous dictum about the historical geographer's need 'to see the land with the eyes of its former occupants, from the standpoint of their needs and capacities' makes clear, he never eschewed imagination; but the main concern was to establish methodological principles operating *within* the boundary of scientific regimentation and not beyond it. Unlike Comte, he did not object to crossing over – there was 'a quality of understanding at a higher plane that may not be reduced to formal process' – but in doing so his phenomenology of landscape remained on the side of positivist science as an unequivocal phenomen*alism*.

Nevertheless, many geographers could not accept that the terms of scientific explanations had to be restricted to the phenomenal level, and on their behalf Hartshorne (1939, 372) claimed that even the most exact natural sciences 'are not concerned merely with the study of objects' but are also engaged in 'the study of forces which cannot be seen or observed directly'. If Newton really had discovered the concept of gravity through an apple falling on his head (which, incidentally, he had not), then 'it was not that force itself he observed by his sense of feeling, but simply the moving apple'. Much the same could be said of geography, he continued, where even a study limited to the visual landscape would consist 'very largely in the study of things not included in the landscape itself', is so far as landscape is 'merely an outward manifestation of most of the factors at work in the area' (Hartshorne 1939, 392–3). While most geographers would presumably regard this as unexceptional, the 'hidden' factors clearly present a problem for the orthodox positivist who, in order to admit them into his conception of scientific investigation, has to introduce a rule of nominalism (Kolakowski 1972, 13) to solve what is in effect the problem of theoretical and observational terms. His answer is to accept that science makes use of abstractions – Hartshorne's 'generic principles' – but to insist that they do not generate any knowledge which is independent of experience, in the sense of allowing access to empirically inaccessible domains of reality. Instead, these theoretical terms have to be unambiguously tied to observational ones by means of correspondence rules (Keat and Urry 1975, 20).

But notice that Hartshorne emphasizes 'generic principles' and avoids mention of causal mechanisms. In the minds of many geographers causality was still uncomfortably close to the discredited thesis of environmental determinism, and Hartshorne was therefore anxious to present geography as a science concerned with 'the functional integration of phenomena' rather than with 'the *processes* of particular kinds of phenomena' (Hartshorne 1939, 593). This reaffirmed its commitment to the discovery of spatial associations: geography was a 'naive science' which looked at 'things as they are actually arranged and related' (Hartshorne 1939, 549). Such a prospectus echoed the empiricist intentions of the German school of history which sought to reconstruct the past *wie es eigentlich gewesen* (as it manifestly was), and this was hardly surprising in an account which relied heavily on an exegesis of the German intellectual tradition. Although geography could now operate at a non-phenomenal level, therefore, it continued to be confined to the demonstration of regularities – or 'simple correlations' – and was not permitted to disclose causalities. This obliged systematic geography to provide clearly defined generic principles, of course, and Hartshorne (1939, 384) conceded that to this extent geography had to be 'a nomothetic science', but its essential task remained the idiographic one of locating these principles in specific regional contexts and describing their interlocking configurations (Hartshorne 1939, 635).

This was subsequently challenged by Schaefer's (1954) attack on what he called the 'exceptionalist' tradition in geography. In an earlier and unpublished manuscript he had defined geography as 'a field inclined and compelled to produce morphological laws rather than process laws', and by morphological laws he said he meant 'patterns' (Schaefer, in Bunge 1968, 19). A similar claim was made in his published essay, and although it still made geography a phenomenalist science, and was in many ways consonant with Hartshorne's own views, the task was now framed in uncompromisingly nomothetic terms.

The dispute between Hartshorne and Schaefer was a bitter and (significantly) a personal one; Bunge (1968, 20) goes so far as to suggest that Schaefer, a refugee from the Third Reich, was the victim of a calculated campaign of persecution which, if not directed by Hartshorne, certainly included him, as well as the FBI, the Office of Strategic Services (which recruited Hartshorne), and the Gestapo. But although Bunge (1968, 12) described the academic dispute between them as 'an argument between Michelson and Newton, or Hegel and Feuerbach', it is extremely difficult to reconcile this with the emphasis of both models on the geometry of the landscape and their common exclusion of any conception of process from geographical inquiry (Sack 1974, 452). Indeed, in a reply Hartshorne (1955, 215) spoke of an 'essential agreement' between himself and Schaefer, and insisted that 'the picture subsequently constructed of major disagreement is therefore false'. It is certainly hard to see how Hartshorne's 'simple correlations' differ from Schaefer's 'morphological laws', given that they can both be reduced to spatial patterns: unless it is in the status which the protagonists ascribe to them. Even then, Hartshorne (1939, 551) had declared that geography seeks to present its knowledge 'in the form of concepts, relationships and principles that shall, as far as possible, apply to all parts of the world' and to organize them into 'logical systems, reduced by mutual connections into as small a number of independent systems as possible'. This is evidently part of the Comtean prospectus, and what Schaefer did was not so much to oppose it as to identify it as the very *raison d'être* of geography, whereas Hartshorne (1939, 644) had taken the view that important as such an enterprise undoubtedly was, it necessarily divested specific regions 'of the fullness of their color and life', which he regarded as the ultimate object of geographical inquiry. *In principle*, therefore, their disagreement was about ends and not means.

For all that, however, it was Schaefer who opened the door for the formal admission of logical positivism into geography. The introduction of the prefix is deliberate, because this was a particular version of the original philosophy formulated by the *Wiener Kreis* (the Vienna Circle), a private seminar conducted by Moritz Schlick in Vienna during the 1920s (see Kraft 1953). One of the members of the group was Gustav Bergmann, who had once taught Schaefer and who, like him, escaped from Hitler's advancing armies to the United States of America. When Schaefer died in 1953 it was Bergmann, by then a colleague on the faculty at Iowa, who undertook the final stages of the publication of his article in the *Annals of the Association of American Geographers*. It is difficult to be precise about its impact: the only immediate reaction to be published was Hartshorne's angry refutation (although it is clear from this that it had provoked much unreported discussion, and it was not until Bunge's own *Theoretical Geography* appeared in 1962, followed by Haggett's *Locational Analysis in Human Geography* three years later, that Schaefer's argument began to be publicly cited as a warrant for the persecution of general laws of spatial organization. Even then it was but one authority among many, so that while Schaefer had opened the door to the premises of logical positivism their admission into the discipline was a slow one and, as it were, through several side-entrances rather than up the main steps.

But we must not lose sight of Bergmann, since his involvement did not end with the publication of Schaefer's essay. Bunge (1962, viii) acknowledged Bergmann's 'especially valuable' assistance in his search for an adequate methodology (although his contribution seems to have been more a matter of ensuring a correct presentation of his former pupil's position than of providing any immediate insights of his own) and in an ambitious attempt to clarify the nature of laws in geography Golledge and Amadeo (1968) used Bergmann's *Philosophy of Science*, which had been published ten years earlier, to argue that Schaefer's endeavours were in fact too restrictive and that a more inclusive typology was possible which would recognize the importance of process as well as morphological laws in geographical explanation. The structure of this typology need not detain us; but their definition of a law as 'a synthetic statement whose validity must be tested by controlled experimentation and/or observation' (Golledge and Amadeo 1968, 762) needs more detailed consideration because it obviously bears on both *la certitude* and *le précis*.

One of the most significant innovations of logical positivism was this very distinction between analytic and synthetic statements. It represented a decisive break with the Comtean model, inasmuch as it accepted that some statements could be validated without recourse to experience. Logical positivism regards such *a priori* propositions as having a truth which arises directly from the meaning of their terms: they are either tautologous ('the earth is a geoid') or self-contradictory, and add nothing 'new' to the corpus of scientific knowledge. These *analytic* statements are not unimportant, even so, because they constitute the domain of the *formal* sciences, logic and mathematics; and these have a strategic role in effecting linguistic transformations and maintaining an internal coherence within the domain of the *factual* sciences, those which treat all other cognitively meaningful propositions as *synthetic* statements whose truth requires some sort of empirical corroboration. The dichotomy was posed in this form by Carnap (1935), but it owed much to Wittgenstein's *Tractatus* and Bergmann (1967, 3) referred to his position 'in the eyes of many' as 'the dominant figure of the movement'. It is only fair to add, however, that their vision was seriously defective; Wittgenstein had intended the *Tractatus* to be interpreted in an opposite sense. It was the statements *outside* the analytic and synthetic sets, those which lie beyond the boundary between 'what we can speak about and what we must be silent about', which he regarded as the most important. As Engelmann (1967, 97) put it, 'positivism holds – and this is its essence – that what we can speak about is all that matters in life. *Whereas Wittgenstein passionately believes that all that really matters in human life is precisely what, in his view, we must be silent about.*'

Nevertheless, it should still be clear that the factual sciences were of primary importance to the logical positivists and that they were firm in their rejection of any metaphysical relativism. Indeed, they not only retained but elaborated a commitment to the *nomological* conception of science, that is, a concern with inferential procedures which determined, in their most general form, *if C then E*. Hempel (1965) maintained that all such deductive–nomological explanations should assume the following schematic form:

$$L_1, L_2 \ldots L_r \qquad \text{(Laws)}$$
$$C_1, C_2 \ldots C_n \qquad \text{(Initial conditions)}$$
$$\rightarrow E \qquad \text{(Event)}$$

in which a particular event (*E*) must *necessarily* follow from the conjunction of the laws (*L*) and the initial conditions (*C*). This implies that the premises, (*L*) and (*C*), can function either as a basis for *predicting* the event or as a basis for *explaining* it. In short, 'prediction and explanation are symmetrical, and deduction ensures the logical certainty of the conclusion' (Harvey 1969, 37).

The deductive–nomological model is made operational by specifying the grounds on which a theoretical proposition can be converted into a law. The normal procedure, of course, is to replace the laws (*L*) by a theory (*T*) from which a set of hypotheses (*H*) can be deduced which, when connected up to the initial conditions (*C*), result in an event (*E'*). In equivalent form:

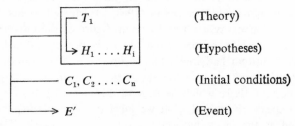

$$T_1 \qquad \text{(Theory)}$$
$$\rightarrow H_1 \ldots H_i \qquad \text{(Hypotheses)}$$
$$C_1, C_2 \ldots C_n \qquad \text{(Initial conditions)}$$
$$\rightarrow E' \qquad \text{(Event)}$$

In order to turn the statements inside the box into a law, a comparison is made between the event predicted by the theory (*E'*) and the event disclosed by empirical observation (*E°*); if they are the same, the hypothesis is accepted and the theory verified.

This is an inadequate foundation for nomological explanation, however, and certainly does not succeed in establishing a law in the Comtean sense. In the first place, it is possible to construct *different* theories which generate the same (empirically consistent) conclusions, and these can only be promoted to the status of laws at the risk of reducing positive science to a cluster of conflicting propositions. Their adequacy cannot be measured by their 'appropriateness' or 'elegance'; commendations like these have frequently found their way into the language of model-building in geography but, as Guelke (1971, 47) points out, 'such terms are subjective and cannot form part of deductive–nomological explanation'. In the second place, it is impossible to guarantee the truth of a statement which, according to Comte, has to be empirically secure *beyond* the necessarily finite range of empirical observations which can be made: 'no matter how many tests are carried out, the law cannot be said to be certainly verified since there always remains the possibility that the $n+1$th observation, following a finite series, will be inconsistent with it' (Giddens 1976, 140).

In the 1930s this formed the starting-point for the 'critical rationalism' of Karl Popper and, as the *Methodenstreit* of the 1960s made only too clear (Adorno 1976), the relationships between him and the logical positivists have remained contentious. Popper himself has an extremely narrow conception of what positivism entails, and has always dissociated himself from its epistemology in general and the Vienna Circle in particular. He regarded the Circle as 'an admirable institution', but insists that the papers which he read in some of the epicyclic groups which formed its halo were essentially critical ones, and that although he counted many of its members as his close friends he never actually joined them (Popper 1976, 80–9; but compare Kolakowski 1972, 209). In some ways the argument resembles the squabble between Hartshorne and Schaefer; critical rationalism can, I think, be accommodated within a wider positivist tradition without too much difficulty. Even so, the differences between logical positivism and critical rationalism *are* rather more substantial than this might indicate.

The heart of the matter was this. Where Schlick (1931, 150; in Popper 1959, 62) had maintained that 'a genuine statement must be capable of conclusive verification', Popper's *Logik der Forschung* (1934) substituted a demand for *falsification*: 'what characterises the empirical method is its manner of exposing to falsification, in every conceivable way, the system to be tested'. The method of the social sciences, he was to say later, consists in 'trying out tentative solutions to certain problems' by attempting to refute them through empirical observation. 'If an attempted solution is refuted through our criticism we make another attempt. If it withstands criticism, we accept it temporarily; *and we accept it, above all, as worthy of being further discussed and criticised*' (Popper 1976a, 89; italics added). Lakatos (1970, 95) hailed this as 'a retreat from utopian standards', one which 'cleared away much hypocrisy and muddled thought', and before his death started to develop a 'sophisticated' version which allows falsification *only* when an alternative ('better') theory has emerged.

This additional requirement, he argued, which was only weakly anticipated by Popper, enables science to advance through progressive 'problemshifts', in which a series of theories is counted as *theoretically progressive* 'if each new theory has some excess empirical content over its predecessor, that is, if it predicts some novel, hitherto unexpected fact', and as *empirically progressive* 'if some of this excess empirical content is also corroborated, that is, if each new theory leads us to the actual discovery of some *new fact*' (Lakatos 1970, 118). Unless the problemshift satisfies both these conditions it is regarded as degenerating rather than progressive.

What these reformulations do, of course, is to qualify, even to 'blow apart' (Lewis, personal communication), the nomological conception of science, and Wilson's (1972, 32) commentary on theoretical geography explicitly recognized the force of the rationalist position: 'Theories are never proved to be generally true. The ones in which we believe represent the best approximations to truth at any one time. To achieve this status, they must be tested and, to date, they must not have been contradicted. We expect, then, that theories will be subject to constant development and refinement.'

Although most geographers would no doubt be prepared to accept this provisional status for their theories and models, very few of them have made any sustained use of Popper or Lakatos (but see Bird 1975; 1977). As a matter of fact, probably the most widely used form of inference in geography – at least in any formal sense – cuts across *both* the nomological model *and* the rationalist response. This is the *inductive–statistical* model, in which one or more probability statements (*P*) replace the laws of the deductive–nomological model and thereby remove the identity between prediction and explanation; in schematic form:

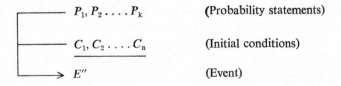

$P_1, P_2 \ldots P_k$	(Probability statements)
$C_1, C_2 \ldots C_n$	(Initial conditions)
E''	(Event)

These statements show the probability of an event of one kind being followed by an event of another kind, and since they are derived from repeated observations of the frequency of the conjoint events any attempt to provide them with a secure logical foundation must lead, in Popper's view, either to an infinite regress or to the doctrine of *a priorism*, both of which were roundly rejected in his original text.

What is more, geography has had a long-standing penchant for other, less formal inductive procedures. One of the first (and most precocious) attempts to qualify geography's commitment to the classical models of Baconian empiricism was von Humboldt's *Kosmos*, but despite its incorporation of elements of Kant's transcendental idealism von Humboldt remained, like so many of his predecessors and his heirs, a strong advocate of the inductive approach (Bowen 1970, 226). Over one hundred years later the quantitative revolution continued – some, Popper notwithstanding, would say strengthened – the hold of inductivism, and still more recently it has received a new impetus from the development of spatial forecasting techniques whose autoprojective methods can be subsumed within the inductive–statistical model (Cliff and Ord 1975, 298; Martin and Oeppen 1975, 95).

This has been the occasion of some concern, not surprisingly, and Wilson (1972, 41) complains that 'there has been far too much emphasis on the use of inductive methods (and associated *statistical* techniques) relative to hypothetico–deductive methods (and a much wider range of associated *mathematical* techniques)'. But, and as we have already seen, this prescription cannot lead to the reinstatement of a Comtean conception of laws in geography, since both inductive–statistical and deductive–nomological models project from a restricted sample on to an unrestricted population and any attempt to construct universal laws must therefore founder on the *general* problem of induction which transcends both specific forms of inference (Harvey 1969, 38).

To summarize, then. Whether geographers have accepted the rationalist critique or whether they have retained their traditional inductivism, they have in the main been ready to circumscribe the status of the laws with which the subject is supposedly concerned: as Harvey (1969, 107) pointed out, 'if we employ very rigid criteria for distinguishing scientific laws, then we can scarcely expect geographical statements to achieve such a status'. But, he continued, 'using less rigid criteria, the identification of laws in geography becomes partly a matter of identifying the relevant theory, and partly a matter of our own willingness to regard geographical phenomena *as if* they were subject to universal laws, *even when they are patently not so governed*' (italics added). The two issues involved here are closely related.

First, the 'relevant theory' has been almost entirely derivative and geography has, for the most part, been totally dependent on constructs developed in other disciplines. Comte's hierarchy of sciences explicitly allowed for this sort of relationship, of course, and the notion of geography as a discipline which pulls together the results of (ancillary) sciences has been a commonplace throughout much of its modern history. Geographers have regularly seen themselves as, if not quite consorts to the Queen of the Sciences, then at least prominent among her courtiers. This allusion to Comte is perfectly apposite, let me say, because the most important sources for contemporary human geography have been neo-classical economics and, to a much lesser extent, functionalist sociology, both of which are predicated on a distinctively positivist epistemology (Hollis and Nell 1975; Keat and Urry 1975). This is not the place to spell out the various connections in any detail, although a detailed exegesis is precisely what is needed because many of the conceptual borrowings have gone unacknowledged and some even unrecognized. But some examples will make the point clear enough for our present purposes, and the discussion can therefore be limited to the founding fathers of location theory, their immediate offspring and some more recent family portraits.

This raises some difficult historical questions, however. In France the spatial dimension had been an insistent element of studies in political economy since the end of the seventeenth century, and this lasted until well into the eighteenth. By the early 1800s, however, it was beginning to falter: 'the abundant flow of ideas integrating the spatial and the economic perspective disappears almost entirely in the nineteenth century like a river vanishing in the desert' (Dockès, cited in Scott 1976, 106). The conjuncture between the eclipse of the spatial tradition and the rise of a Ricardian political economy is clear enough; but what is more important in the present context is that it also coincided with the *rise* of Comtean positivism. In France, therefore, a lacuna appears between the early growth of the regional sciences and the development of the epistemology to which they subsequently subscribed. In Germany, widely regarded as the traditional home of location theory, this becomes a *dis*juncture; the regional sciences developed throughout the nineteenth and early twentieth centuries in an intellectual milieu which was in fact strongly *resistant* to positivism and its naturalist pretensions. But despite this somewhat peripheral, almost cloistered existence location theory was readily assimilated to the positivist paradigm.

J. H. von Thünen's *Der Isolierte Staat* (1826) depended on a classical conception of rent to describe its concentric circles of rural land use around a central city, but it contained an unequivocal legitimation of marginalism (Barnbrock 1976, 200), and its extension into urban land use theory, notably through the efforts of Alonso (1965), Muth (1969) and Mills (1972), was accompanied by an easy translation into the neo-classical framework. Alfred Weber's theory of industrial location presents rather more ambiguities. He began his work at a time when German idealism was still in the ascendant, and he developed his location theory in response to two of its intellectual programmes. The first of these was a continued commitment to the *Geisteswissenschaften*, the 'human sciences', on the part of the universities: they had detached themselves from the narrowly material concerns of the new bourgeoisie and were engaged in the illumination of a purely German *Geist* which they saw emerging out of the precariously reunited fragments of the Empire. Weber was ready to recognize the importance of these efforts, but he wanted to know whether it was 'sensible for us to argue about cultural and social motives when perchance we are simply fettered by the iron chains of hard economic forces?' To this extent, his location theory represented a qualification of the profound anti-naturalism and anti-materialism which dominated contemporary discourse, and as such his search for 'pure laws of industrial location' which would be 'independent of any particular kind of economic system' was clearly consonant with the demands of positive science. But at the same time Weber was greatly affected by the activities of his elder brother, Max. He was impressed both by the claims of *verstehende* social science itself (see below, pp. 131–4), which challenged the epiphenomenal status which materialism then accorded the cultural sphere, and also by Max's programmatic interventions in the *Methodenstreit*, which established the importance of historical specificity to a properly constituted science of society.

These considerations were certainly *not* consonant with the universalist aspirations of positivism, and they prompted him to qualify the abstract formulation of location theory set out in his *Uber der Standort der Industrien* (1909) and to argue that the contemporary pattern of industrial development could not be explained by the 'pure' laws of location alone and that it resulted 'to a large extent from very definite central aspects of modern capitalism which might disappear with it'. But Weber only ever sketched out a theory of capitalist location in a brief essay in the *Grundriss der Sozial-*

ökonomik which was never translated into English, and the foundations of his 'epistemological break', if we can call it that, did not receive a full discussion until his *Kultursoziologie*, where they were easily overlooked by later generations of geographers and regional scientists (Gregory 1977). Subsequent reformulations, from Palander (1935) through Hoover (1948) and Isard (1956) to Nijkamp and Paelinck (1973), therefore had little difficulty in fastening on the more accessible, which is to say the more positive, aspects of Weber's project, and his 'pure' transport-minimization model was plucked from the world of locational triangles and Varignon frames and placed firmly in the equally abstract world of neo-classical economics. In contrast to the interpretative violence done to Weber, however, the other major contributions were received in a much more straightforward manner. Walter Christaller's *Die Zentralen Orte in Süddeutschland* (1933) posed very few problems, and present-day analyses of the revealed space-preference structures of consumers in central place systems are little more than translations of neo-classical utility theory into a spatial context (Rushton 1969; Clark and Rushton 1970; Girt 1976). Similarly, although August Lösch's *Die Räumliche Ordnung der Wirtschaft* (1940) contained no explicit account of supply and demand schedules it represented one of the first attempts to derive a spatial equivalent of the general equilibrium theory of neo-classical economics: and it was by no means the last.

There ought to be no need to labour the point any further: enough has been said to disclose the size of the intellectual overdraft. It meant that once geography began to draw on neo-classical economics in order to articulate more formal theories about the space-economy its own, somewhat fuzzy empiricism was considerably strengthened and sharpened. As Harvey (1972, 325) conceded, logical positivism (and its variations) brought 'hard-headedness' to geography and, as the second half of his prescription for laws in geography made clear, this was not going to be damaged by repeated banging against one wall after another to discover potentially refuting observations on the other side. If it was not possible to show that geographical phenomena were subject to universal laws, there was still some value in regarding them *as if* they were. Injunctions of this sort flow from an instrumentalist conception of science.

Instrumentalism regards theories as devices whose utility is at stake; their 'truth' cannot be an issue since no conclusive validation can be provided for them, and so science is justified in adopting a more pragmatic set of standards in which its propositions are evaluated according to the success of their predictions and nothing else. Whether their assumptions are in some way 'invalid' or 'unrealistic' is irrelevant: it is the end result which matters. Instrumentalism plays an important supporting role in neo-classical economics, and so it is not surprising to discover that it has been carried over into much of modern geography, where its emphasis on 'goodness of fit' has had two consequences of major significance.

First, it has allowed an extremely narrow, even superficial, formulation of 'spatial process' to emerge, in which empirical space–time variations are made to conform to a typology of corresponding forecasting models (see Martin and Oeppen 1975; Cliff, Haggett, Ord, Bassett and Davies 1975; Haggett, Cliff and Frey 1977). This is frequently helpful, of course – it would be churlish to say otherwise – but the empirical identification of appropriate model structures ought not to become a substitute for the proper specification of the mechanisms involved. It would, moreover, be foolish to deny that this has been recognized: Haggett, Cliff and Frey (1977, 517) have suggested that 'the ability to forecast accurately should represent an ultimate goal of geographical research' precisely *because* 'this ability ought to imply a fairly clear understanding of the processes which produce spatial patterns'. It ought, certainly; but all the time that an instrumentalist definition of process is accepted progress is unlikely to be rapid. Instrumentalism is simply not concerned with these kinds of endeavour at all. When Harvey (1973, 176) asks 'what it is that makes micro-economic theory so successful (relatively speaking) in the modelling of urban land-use patterns, when it is so obviously wide of the mark when it comes to modelling the real processes that produce these patterns', the answer is, as Ive (1975, 22) pointed out, that 'it would be surprising if neo-classical theories did appear inadequate in terms of their own criteria of statistical testing since they are mainly designed solely with a "good fit" in mind'.

Secondly, and connected to this, instrumentalism has promoted a limited, at times almost an opportunist, image of geography as a policy science. When, for example, Bennett used transfer function (TF) and autoregressive-moving-average (ARMA) models to replicate space–time variations in unemployment in the north-west of England, he concluded that 'whilst the real world certainly does *not* behave as a low order TF or ARMA model, it *can* be treated as such' (Bennett 1974, 172), even that it ought to be, since this would permit 'the simulation of the efficacy of application of past policy instruments, and will facilitate the formulation of policy since the range of possible outcomes is known' (Bennett 1975, 887). And this is *not* a carefully chosen and isolated instance: both Olsson (1972) and Lewis and Melville (1977) have shown that the instrumental approach of the social engineer dominates geography and the other regional sciences *in general*.

In sum, then, and generalizing from this discussion, if instrumentalism appears to compromise the apodictic certainty and the supreme truth required by positivism, at least as Comte understood it, these two consequences show, in turn, that however much it might violate the classical ideal of *le précis*, instrumentalism is nevertheless predicated on an empiricist conception of knowledge (*le réel*) and holds to a technical conception of science (*l'utile*): in short, that the assumptions and aspirations of instrumentalism and positivism are very closely bound up with one another (Keat and Urry 1975, 65), to the extent that, in geography anyway, they have become virtually indistinguishable.

The technical conception of science involves more than the divorce of 'means' from 'ends', however, more than a distinction between 'policy-relevant' and 'policy-forming' contributions; it also implies that 'we understand a state of affairs only to the extent that we have the knowledge of what to do in order to control it' (Fay 1975, 40). Perhaps its most powerful and certainly its most explicit expression has been the claim that geography ought to concern itself with the analysis of *control systems*.

Chorley and Kennedy (1971) have suggested that the various kinds of system can be arranged in a hierarchy of four levels:

(1) *Morphological systems*, which consist solely of the physical properties of their components and where the relationships between them are expressed through a web of statistical correlations;

(2) *Cascading systems*, which consist of a chain of sub-systems linked by a cascading throughput such that the output of one subsystem forms the input for the next;

(3) *Process-response systems*, which are formed by the intersection of morphological and cascading systems;

(4) *Control systems*, which are process-response systems which are structured by the intervention of decision-making agencies at certain key points ('valves') so as to alter the disposition of the

throughputs in the cascading system and hence change the equilibrium relationships in the morphological system.

Although both environmental and social systems can be modelled in terms of morphological, cascading and process-response systems, it is clear that these first three levels are regarded as the primary domain of the earth sciences and that the proper focus for 'a geography without adjectives' (Anuchin 1973, 62) is the fourth level. There, particular attention will be paid to 'the equilibrium relationships of the process-response systems and the manner in which they can be disturbed by inadvertent human action leading to a degradation of earth resources, or thoughtfully regulated so as to exploit their inherent operational characteristics as part of wider geographical systems being controlled for the well-being of a wide range of living things, man included' (Chorley and Kennedy 1971, 343).

The priority which this gives to environmental management represents a polemical attempt to heal the breach between human and physical geography, and it portrays the relationship between man and nature as that of 'an increasingly-numerous, increasingly-powerful and progressive, if capricious, master and a large, increasingly-vulnerable and spitefully-conservative serf' (Chorley 1973, 157). It is no accident, then, that man is at the top of the hierarchy, Prospero to nature's Caliban, and that the constraints on his thoughts and actions are essentially environmental ones. Comte's hierarchy of sciences also recognized physical constraints at the lower levels, but his primary concern was with a set of irreducible and intrinsically *social* constraints. But since the systems typology can characterize society just as easily as it can the environment this is no problem, although since 'the operation of control is forbiddingly complex in systems of the scale normally studied in geography' (Langton 1972, 109) the discipline clearly has a long way to go before it even approaches the rigid formalization of, say, Talcott Parsons.

According to Parsons, all social systems display four functional exigencies: they must adapt to fluctuations in their operating environment; they must define and attain specific goals; they must maintain their integration; they must resolve their latent tensions. The same applies to individual sub-systems, as indicated in Figure 1, so that considerable interest is attached to the way in which sub-system adjustments are mediated in order to secure the survival of the system as a whole. Control, then, is absolutely central to Parsons's 'problem of order' (Giddens 1976, 98), and in much the same way Chorley (1973, 161) argues that systems can be restructured through a process of active control 'involving the impelling of systems on time-trajectories through sequences of states, each different, probably non-recoverable, and *presumably* ever more adapted to the evolving needs of man in society' (italics added).

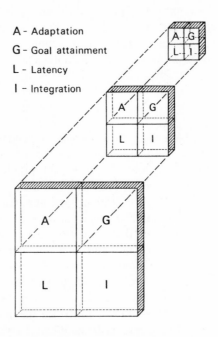

A - Adaptation

G - Goal attainment

L - Latency

I - Integration

Figure 1 *Talcott Parsons's conception of social system*

The presumption is considerable; there is no Hegelian *Aufhebung* in Chorley's vision of the master–serf (man–nature) relationship (see below, pp. 153–4), and the object of the control process is to preserve rather than transcend the existing system. Chorley (1973, 157) is right to observe that 'it would be a foolish master who did not diligently study the characteristics of his subordinate in order to so moderate his actions as to extract the maximum efficiency from his employee and to keep him fit for future work', but this is not simply an analogy: the interchangeability of systems analysis makes this apply to real masters and serfs as much as to metaphorical ones. Those who defend systems analysis against socialist critiques by citing its extensive application in Soviet Russia are not only guilty of double standards, therefore; they also fail to recognize that, as a matter of historical record, 'the [positive] scientific method led to the ever-more-effective domination of nature [and] thus came to provide the pure concepts as well as the instrumentalities for the ever-more-effective domination of man by man *through* the domination of nature' (Marcuse 1972, 130).

In case this is misunderstood, the sincerity or otherwise of advocates of systems analysis is not at issue; rather, the positive conception of science to which they subscribe *in itself* entails a particular conception of practical life, irrespective of what their own intentions might be, and the connections between the two are determined by what Habermas (1972, 309) describes as a deep-seated *cognitive interest* in technical control (see below, pp. 68–70; 157–60).

Having said that, the conscious wish to inform policy decisions was undoubtedly responsible for much of the initial enthusiasm for the systems approach in geography. But this was reinforced, and at times almost overridden, by the simultaneous realization that 'the progress of science might be more efficient if some basic theory could be derived which covered isomorphisms between systems, and if criteria could be developed to facilitate the transference of ideas and formulations across disciplinary boundaries' (Harvey 1969, 470). This would allow what Ackermann (1963, 438) called a 'collaborative definition' of research priorities and procedures, in the course of which particular fragments would gradually join together in a general system of knowledge which transcended the specificity of existing disciplines and sub-disciplines. The ideal has a long history in geography, of course; von Humboldt's *Kosmos* was one of the earliest modern attempts to provide a theoretical conception capable of systematizing scientific knowledge about man and nature in geographical terms, and for all its Kantian trappings it spoke directly to Comte's vision of *le relative*. So too, in its way, does General Systems Theory, described by von Bertalanffy (1973, 31) as a 'new discipline' concerned with the 'formulation of those principles which are valid for "systems" in general', and it was on this basis that Warntz (1973) – among others – argued that geography ought to regard itself as General Spatial Systems Theory.

The attraction of this is that, like systems analysis, it appears to offer a way of resolving the dualism between human and physical geography; what is distinctive about it, however, is that it claims to do so 'by formulating theoretical statements of the [common] properties of different types of systems' (Conacher 1969, 157). Applications of the law of allometric growth to urban systems, fluvial systems and volcanic systems (Nordbeck 1965), the use of entropy concepts to characterize the spatial organization of both human and physical landscapes (Leopold and Langbein 1962; Chapman 1970; Medvedkov 1970; Connelly 1972; Marchand 1972; Batty 1974; Chapman 1977), and the analysis of space-filling processes common to central place systems and drainage basins (Woldenberg and Berry 1967; Woldenberg 1970; 1972) all indicate that this proposal has secured at least a measure of support. This groundswell, if such it is, might appear still more formidable if the constructs shared by human and physical geography which were not originally derived from systems theory but which can be translated directly into its vocabulary are counted in. Jennings (1973, 124) notes that 'most of the central ideas of general systems theory are without question valuable, but most originated in or were applied to geomorphology without reference to systems theory', and Langton (1972, 159–60) similarly concludes that 'many of the concepts of systems theory are already used in [human] geography without the attendant jargon and without apparently drawing direct inspiration from the literature of systems theory'.

This manoeuvre is not intellectual gerrymandering, because a formal acknowledgement of systems theory would presumably encourage the integration envisaged by Comte's final precept. Indeed, the difficulty which Conacher (1969, 157) sees in General Systems Theory is totally misplaced: he recognizes, quite correctly, that the unity of positive science ought to be pre-eminently a methodological one, deriving from the way in which scientific inquiry is conducted 'rather than from the common properties of sets of diverse phenomena', but *le relative* clearly allows for this to be buttressed by clusters of common constructs.

But other misgivings have been voiced, and not only by those who remember the moon General Systems Theory once promised and now sniff disparagingly at the green cheese which they see offered in its place. To reconstruct geography in these terms is, for many, at once too inclusive and too restrictive an enterprise. Chorley (1973, 158) has insisted that while geography *is* 'an inherently spatial discipline' it is by no means the only one, and that its ultimate and distinctive concern must be with the spatial manifestations of the interaction between man and his material universe. Although the equivalence of spatial structures in physical and human geography 'manifestly enriches the two disciplines' – see, for example, Haggett and Chorley (1969) – it cannot, in his view, provide 'the kind of fusion between physical and human geography which is needed' (Chorley 1971, 96). He maintains that this has to come, as we have seen, from systems analysis and not from General Systems Theory. Systems analysis is 'an empirical method' (Langton 1972, 170; Chisholm 1967, 49) and as such it can interrogate the spatial inter-sections between regionally-specific physical and human systems. Moreover, it circumvents Conacher's problem by restoring method-ological primacy, and can therefore, I think, be located at the heart of a multi-dimensional matrix formed by *le réel, la certitude, le précis, l'utile* and *le relative.*

The hold of systems analysis on contemporary geography is diffi-cult to assess, but three of the most recent – and certainly the most innovative – positive texts to be published (Haggett, Cliff and Frey 1977; Chapman 1977; Bennett and Chorley 1978) are all couched within its problematic. Together, they seem to fulfil Stoddart's (1967, 538) prediction, made ten years earlier, that systems analysis 'at last provides geography with a unifying methodology, and using it geography no longer stands apart from the mainstream of scientific progress'; instead, it is brought 'back into the realms of the natural sciences' . . .

The materials assembled in this chapter indicate that geography embraced a philosophy which, if not entirely 'new' certainly became more 'positive' as time went on. These twin designations are taken from Harvey (1969, 486), but nowhere else in his discussion of *Explanation in Geography* is positivism mentioned as such. This omission is in part a result of the distinction he makes right from the start between methodology and philosophy, one which he has since admitted to be untenable (Harvey 1972, 324), but I think more importantly it is also because geography has (with some notable exceptions) paid scant attention to its epistemological foundations. This means that I do not for a moment pretend that the incorpora-tion of positivism into geography was necessarily a considered one, much less claim that it derived directly from Comte himself. There can clearly be many reasons for accepting a particular doctrine, and self-awareness is not always at the top of the list. But what I do contend is that modern geography can be related quite explicitly to that original prospectus, even allowing for the variations introduced by logical positivism, critical rationalism and instrumentalism, and, further, that it is *necessary* to do so, for two reasons.

In the first place, geography *has* become more sensitive to epistemological considerations in recent years, but this has all too often amounted to the use of 'positivism' as a term of blanket condemnation rather than of careful identification (Cooke and Robson 1976, 85). That geography has not been alone in its horrified recoil from the unacceptable face of positivism is scarcely a comfort. It is admittedly not easy to characterize positivism in a simple and unambiguous set of propositions – the details in my narrative present little more than a caricature of even its most striking features – but it requires something more than a casual dismissal of one position to justify the adoption of another.

This, of course, is the second reason. By relating geography to Comte's five precepts and bringing the hidden assumptions of positivism out into the open, I hope that the critique which follows will be both more incisive and more convincing. I have attempted in my presentation of positivism, no doubt not entirely successfully, to retain a proper distance from what I believe to be its improper foundations and its provocative implications. The next chapter is intended to make up for this reticence.

2 In place of spatial science

It all began with a suspicion (perhaps exaggerated) that the Gods did not know how to talk. Centuries of fell and fugitive life had atrophied the human element in them. . . . Suddenly we sensed that they were playing their last card, that they were cunning, ignorant and cruel like old beasts of prey and that, if we let ourselves be overcome by fear or piety, they would finally destroy us.

We took out our revolvers (all of a sudden there were revolvers in the dream) and joyfully killed the Gods.

Jorge Luis Borges: *Labyrinths*

Borges's words almost invite the re-enactment of Alvin Gouldner's *The Coming Crisis of Western Sociology*, published in 1970, which is prefaced by Nietzsche's 'Here are the priests; and although they are my enemies . . . my blood is related to theirs'. The comparison is worth pursuing, even if it has to be rejected in the end, because it provides a context for the discussion which follows. There are three points I wish to make here.

First, Gouldner was writing at a time when, as he says himself, 'the old order had the picks of a hundred rebellions thrust into its hide' and when the social sciences theorized 'within the sound of guns'. The experience was not unique to sociology and the drums of social concern were beginning to beat in geography at more or less the same time. A conference of the American Sociological Association at Boston in 1968 had been paralleled by a shadow series of unscheduled meetings organized by a radical caucus, which culminated in Nicolaus's angry denouncement at the plenary session of those who watch over 'the inequitable distribution of preventable diseases, over the funding of domestic propaganda and indoctrination [and] over the preservation of a cheap and docile labour force'. 'The professional eyes of the sociologist are on the down people,' he declared, 'and the professional palm of the sociologist is stretching toward the up people' (Nicolaus, cited in Gouldner 1970, 10).

A conference of the Association of American Geographers in the same city only three years later revealed that

> many geographers were deeply frustrated by a sense of failure, conscious that the knowledge they already possessed was not being put to good use, that much had been learned about ways and means of reducing hunger, disease and poverty, but little had been achieved, that educated people had not been instrumental in stopping a barbarous war [in Viet Nam], and that, within their own Universities, they had failed to bring about overdue reforms [Prince 1971, 152].

It is all too easy to be seduced by casual comparisons, of course, and we should not lose sight of the radical dimension which Reclus and Kropotkin added to European geography at the turn of the twentieth century (Stoddart 1975; Peet 1975; Galois 1976), but even with these qualifications the convergence remains a striking one. Smith (1971, 157) concluded from the AAG conference that the vaunted New Geography of the 1950s and 60s had grown old, and that the heightened social awareness within the discipline would necessarily diminish the force of its positivist orthodoxy. But however appropriate this may have been then, it is hardly sufficient today. The social crisis has become much more apparent since those meetings (and the private murmurings which they publicly crystallized), and in the wake of a severe and protracted economic recession the academic social sciences – including geography – have been obliged to incorporate some kind of critical analysis into their syllabuses, if not always into their prescriptions. In short, the conjuncture is now different, and a direct translation of Gouldner's thesis into geography ought to be resisted on these grounds alone. Having said that, though, the critique of positivism remains an urgent task if the critical potential of the discipline is to be realized rather than institutionalized, in both senses of both words: that is, if the nature of critical theory is to be properly recognized and its imperatives achieved in social practice, and if it is not to become just one more compartment of a subject already fragmented into too many subdisciplines, sustained in an artificial form and eventually discarded because it has outlived its original purpose.

Secondly, and again as Gouldner recognizes, sociology had become part of an emergent popular culture in the period after the Second World War in a way which, for a variety of reasons, was denied to geography. This was not without its drawbacks and, as Giddens (1976, 15) says, many concluded from this – with some

justification – that the 'findings' of sociology, once they were deci-
phered, told them nothing which they did not already know. This
frank confession of failure will no doubt reinforce the prejudices
which some geographers still have about sociology, ones which have
not changed all that much from the immediate post-war years when
the Education Committee (*sic*) of the Royal Geographical Society
could dismiss social studies in schools as the result of 'what happens
when a lemon is squeezed : the juice is removed, and only the useless
rind and fibres remain' (cited in *Geography*, 1950, 181). They would
do well to reflect on the likely outcome of a public scrutiny of their
own 'findings', had they been rescued from their relative obscurity.
But the point is that they were not, and to that extent there was a
sense in which sociology was more intimately involved with society
at large, a sense in which its perspectives (where they were not
rejected out of hand) served to increase political literacy and to
inform political action in a particularly direct way. There was, then,
and for that matter still is, considerable 'slippage' (Giddens 1976)
between technical and lay discourse, and yet for all the geographers
who came to see their own discipline in a light far removed from the
watery glow of the explorer's hurricane lamp, it still retained, in the
public mind at least, much of that pith-helmet and puttees image
which was to become increasingly outdated as time went on.

   These comments are not meant to denigrate the much more recent
achievements of the American High School Geography Project, the
alternative and the revised syllabuses of English examination boards,
or the Geography for the Young School Leaver Project of the Schools
Council; but however exciting these developments and others like
them may be for the teaching profession they have yet to make a
similar impression on a wider audience. It is part of my argument
that they are intrinsically unlikely to do so because, to put it
obliquely, even if they succeed in substituting a different image they
are still using the same camera : they remain committed to a positivist epistemology which makes social science an activity performed
on rather than *in* society, one which portrays society but which is
at the same time estranged from it. I want to leave this as an unsupported assertion for the moment, but it is relevant here even as
it stands because it makes sociology's active participation in popular
culture much, much more than a mid-twentieth-century trivializa-
tion of a serious discipline and, by the same token, turns geography's
continuing cultural isolation into an object of concern rather than
congratulation. Gouldner's discussion is at once made inappropriate

for a geographical audience, while the importance of a sustained critique of positivism is confirmed.

Finally, and most obviously, Gouldner diagnoses a 'coming crisis' in sociology. For its part geography has also been made uncomfortably aware of the impending Ides of March. The disillusion with the quantitative revolution has already been noticed (see above, pp. 20–1), and a report on geography in the United Kingdom between 1972 and 1976, prepared for the 23rd International Geographical Congress in Moscow, disclosed 'a frustrating anticlimax after the heady enthusiams and elevated hopes' of capturing 'the great white whale of a general spatial science' (Cooke and Robson 1976, 81). If a moratorium on Moby Dick has yet to be declared, the whalers are nevertheless drifting in an empty and increasingly stormy sea, their harpoons rarely puncturing anything larger than a plastic bottle. And yet to speak of a crisis in geography may be to exaggerate the situation, not because, as Chisholm (1975, 173) contends, a counter-revolution would be misconceived, but because, as Kuhn's (1962; 1970) account of paradigm change indicates, it can be argued that science progresses through a revolutionary interruption of normal practice, which would make any 'crisis' a transitory rather than a terminal condition, eventually to be resolved by a transformation of the existing paradigm. But this is an inadequate response on at least three counts.

First, although Kuhn's model has been widely accepted by polemicists in geography it has not been greeted so favourably by philosophers of science (Stoddart 1977). The most important qualification as far as the present discussion is concerned is that Kuhn ignores the relationships between science and society (Harvey 1973, 121), since it follows from what has been said already that these must be of central importance in accounting for any crisis. It was the manifest inability of existing theoretical constructs to explain contemporary events that obliged the traditional sciences to recognize some of the claims of critical theory, simply to retain their social credibility.

Secondly, even if Kuhn's analysis is accepted it can only promise a resolution of the crisis at some future date and obviously cannot prescribe one now. It is one thing to treat crises as recurrent events – Chorley and Kates (1969, 1) suggest that 'it is of the nature of scholarship that all scholars should think themselves to be living at a time of intellectual revolution – but quite another to come to terms with the difficulties which underlie them. Kuhn can offer very little guidance on this score, apart from an extremely contentious proposal for the institutionalization of paradigms (Feyerabend 1970; 1975).

Thirdly, in so far as geography may properly be said to be in a state of crisis, this is in many respects a general crisis which has struck at the heart of all the social sciences and, according to Husserl (1970), at the heart of science *tout court*. Husserl's contribution is particularly significant since it attempts to return science to society through a rejection of the objectifying aspirations of the modern scientific enterprise, and this *general* project is so clearly predicated on a dismissal of positivism that it can be aligned with concurrent developments in the *individual* social sciences. This sort of global perspective is in contrast to the much narrower context in which Gouldner conceives of a crisis (Smart 1976, 10–39), but it is the most appropriate vantage-point from which to view the state of geography today and to see just how deep and pervasive are the roots of its current malaise.

To summarize. These three elements – conjuncture, culture and crisis – suggest that Gouldner's work does not provide a ready template from which to trace the outlines of the dilemma now facing geographical inquiry. But they also make it clear that the critique of positivism has to occupy a pivotal role in our future discussions. This is scarcely a novel suggestion, of course; it would certainly not be true to say that all geographers greeted positivism with open arms or that those who did do so have necessarily retained their allegiance. But the early stages of the debate were focused on a particular methodological *form* (the quantitative revolution) and not on its deeper and more enduring epistemological *structure*, notwithstanding the repeated declarations that something more than mere numeracy was involved (for example, Burton (1963); Harvey (1969)). The later stages have been more productive, however, and perhaps even constitute a *Methodenstreit* of their own: the rest of this volume attempts to bring together and to extend their arguments, and to explore their implications.

The discussion which immediately follows is more limited. It re-examines each of Comte's precepts and their variations in turn, in order to isolate, in a very preliminary fashion, some critical reactions which can be pursued in greater depth in subsequent chapters. It may be as well to admit right now that these reactions are in themselves problematic, since they are characterized by their own shortcomings and schisms and are not entirely consistent one with another. They do not amount to a ready-made route out of the waste land, therefore, and it would be dangerous to pretend that they did: if anything, to continue the metaphor, they represent a set of faint tracks and traces, crossing, joining, diverging, even vanishing; nothing more. But it still ought to be possible to identify a starting-point and a set of possible destinations, even if we are as yet unable to find the shortest and safest path between them. To repeat: its discovery will not be easy or accidental, much less automatic, and geographers will not be able to stumble blindly along corridors cut by practitioners from other disciplines. The construction of an adequate conception of critical science cannot be a passive exercise.

What this means – and this will become increasingly obvious as my presentation unfolds – is that it is *impossible* to conduct the search by relying on a purely geographical literature (even supposing we could ever agree on what this is). Wherever possible I have in fact tried to relate the constructs involved to their counterparts in geography, but this is only to make a closely textured argument more accessible and, in any case, it frequently happens that constructs of a comparable power – and this often implies constructs of a comparable difficulty – have not been developed within geography at all. If the questions they ask and the answers they provide lie outside the established problematic of geography, however, this does not entitle us to ignore them or to bury them in an impressive series of footnotes. As Wise (1977, 4) has said, 'if we are fully to understand the significance and role of our own work and to measure progress in geography, we must attempt, more generally than we do, the difficult task of placing the problems that we select for study in the widest possible scientific and philosophical context'. And in doing so we cannot seriously expect our view of those problems, ultimately our conception of geography itself, to remain undisturbed.

A re-examination of the precepts of positivism

The most fundamental component of positivism, whether in its classical or its modern versions, is the commitment to empiricism established through the rule of phenomenalism. Empiricism, from the Greek *empeiria* (experience), emphasizes the strategic role of scientific observations, and makes two assumptions about the language in which they are made (Keat and Urry 1975, 19). These are:

(1) that the observation language is *ontologically privileged*, which simply means that observational statements are the only ones which make direct reference to phenomena in the real world; the corollary of this is that theoretical statements are not privileged, and since they are unable to provide access into some 'other' realm then ultimately they too have to refer to phenomena in the real world, but they do so indirectly via 'correspondence rules';

(2) that the observation language is *epistemologically privileged*, which simply means that observational statements can be declared true or false without reference to the truth or falsity of theoretical statements.

In short, these assumptions amount to the claim that although science relies on a theoretical language *and* an observation language the first of these is somehow parasitic upon the second and 'probably ought to be eliminated from scientific discourse by disinterpretation and formalization, or by explicit definition in or reduction to observation language' (Hesse 1974, 9).

I will now argue that both of these assumptions ought to be abandoned.

First, ontological privilege can be challenged by the counter-claims of realism. *Realism* regards the positivist view of explanation as seriously deficient because it confuses relations of logical necessity with relations of natural necessity (Keat and Urry 1975, 27): as we have seen already, 'deduction ensures the logical certainty of the conclusion' (Harvey 1969, 37) but cannot of itself specify the causal mechanisms through which this is brought about. In a different way Hartshorne's rejoinder to Sauer spoke to the same point in arguing for the incorporation of 'hidden factors'; but what was missing, in the wake of the determinist controversy, was a willingness to treat causality in anything other than Humean terms (that is, through anything more than statements in which events or observations of one kind are regularly followed by events or observations of another kind: 'simple correlations').

Realism allows for more than this by establishing three fundamental principles which collectively overturn empiricism (Harré 1972, 91):

(1) some theoretical terms can be used to make reference to hypothetical entities;
(2) some hypothetical entities are 'candidates for existence';
(3) some candidates for existence are demonstrable.

The effect of these claims is to allow *theoretical* statements to make ontological commitments, which means that phenomena can legitimately be explained by recourse to mechanisms (or 'structures') that, typically, are beyond observation.

Moreover, realism insists that these mechanisms disclose natural necessities: 'each state of the world must be as it is' in the sense of 'being the necessary result of a necessary cause' (Hollis and Nell 1975, 68), and this collapses the positive distinction between analytic and synthetic statements. 'No hypothesis is finally synthetic or contingent, since it is in the end either true of a necessary order [and hence "analytic"] or false and so contradictory' (Hollis and Nell 1975, 69): 'finally' and 'in the end' because realism still accepts that hypotheses must be *tested*. Some of the problems which this presents can be seen from an examination of the second assumption.

Secondly, then, epistemological privilege can be challenged by a 'network model' of scientific inquiry which is broadly conformable with the realist position. The account of it which follows is taken from Hesse (1974), but it was first presented in Colodny (1970) and is an extension of Duhem (1954; 1969) and Quine (1961).

Hesse (1974, 11) starts from two premises:

(1) 'All descriptive predicates, including observation and theoretical predicates, must be introduced, learned, understood and used, either by means of direct empirical associations in some physical situations, or by means of sentences containing other descriptive predicates which have already been so introduced, learned, understood and used, or by means of both together;

(2) 'No predicates, not even those of the observation language, can function by means of direct empirical associations alone.'

This maintains what might be called a *reflexive* relation between theory and observation, the one mediating the other.

I want to be quite clear about what this entails. *All* descriptive predicates are supposed to be (ultimately) grounded in physical situations, so that in contrast to Hempel's (1952, 36) characterization of theoretical predicates as knots in a net, tied together by definitions and theorems, and the whole mesh 'floating "above the plane of observation" to which it is anchored by *threads of a different kind*, called [correspondence rules], *which are not part of the network itself*', Hesse's argument is that the mesh 'is not floating above the domain of observation; it is attached to it at some of the knots. *Which* knots will depend on the historical state of the theory and its language and also on the way in which it is formulated, and the knots are not immune to change as science develops. It follows, of course, that [correspondence rules] disappear from this picture' (Hesse 1974, 27).

But if this reflexive relation does not separate theory from observation as positivism tries to do, it does not confound them either: in terms of our earlier analogy, the network model does not reverse host and parasite. To do so would be to move from realism into *conventionalism* and to assume:

(1) that theories determine observations rather than provide different ways of structuring the same ones;

(2) therefore proponents of different theories see the world in different ways;

(3) as a result, comparison of different theories through common observations is totally impossible.

This would make science an autistic enterprise, with each scientist 'isolated in his own system of meanings as well as within his own universe of observed things' (Scheffler 1967, 17). Individuals could move from one prison to another, but only by finding their way through an elaborate maze.

Feyerabend (1970; 1975) would probably prefer to say that this ought to make science an anarchistic enterprise in which, as he certainly says, 'anything goes', but realism regards science as consisting 'in some sense in the permanent and cumulative capture of true propositions corresponding to the world' (Hesse 1974, 290) – its propositions are supposed to be true by necessity not by convention – and so it cannot base itself on the theory–observation relationship set out in (1) above.

The solution which Hesse (1974, 33) proposes is to make a distinction between those observations which are theory-*determined* and those which are theory-*laden*. This is not a semantic artifice, but an attempt to rule out theory-determined observation statements, whose terms presuppose the truth of the theory under test, and to say that the meanings of the theory-laden observation statements which remain are not guaranteed by, as it were, the tightness of individual knots but by the strength of the mesh as a whole or, more formally, its coherence. This means that 'the situation of crucial test between theories is not correctly described in terms of "withdrawal to a neutral observation language" ' – the dichotomy is fully exhaustive and observations *must* be either theory-determined or theory-laden – 'it should rather be described as exploitation of the area of intersection of predicates and laws between the theories' (Hesse 1974, 35).

Olsson (1971) has attempted to devise a similar model for the conduct of geographical inquiry, which recognizes that 'all observations are more or less theory-laden' so that the idea of 'sharp distinctions between theoretical and observational languages is dubious': instead, there has to be a 'freer interplay' between them. Although, unlike Hesse, Olsson retains the correspondence rules they are interpreted in a much more catholic way than positivism would allow and are, I think, close to the knots linking theoretical and observational domains in the network model. Thus he says that 'we frequently convey meaning on theoretical terms not only by relating them to observables but also by relating them to antecedent theories which typically contain references to both observables and unobservables'; 'correspondence rules and reduction functions take on pivotal roles in the games of scientific progress; each term has a primary sense specified by the initial theory, and a secondary sense, which is ascribed to it through the couplings provided by reduction functions and correspondence rules' (Olsson 1971, 550).

These materials are not identical, of course, but the argument which they begin to suggest, however incompletely, is one which represents scientific discourse as a structure in which theories and observations progressively transform one another. This relationship is contrasted with those contained by empiricism and conventionalism in Figure 2.

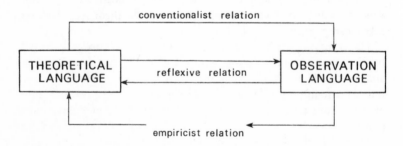

Figure 2 *Theory–observation relationships*

Hesse's grounding of the theory–observation relationship is not wholly successful, however, and for two main reasons. In the first place, although the privileges which empiricism accords to the observation language have to be rejected – as Popper (1959, 59n) says, observation is always observation in the light of theories – as he also says, conventionalism is 'a system which is self-contained and defensible. Attempts to detect inconsistencies in it are not likely to succeed.' He rejects it in the end, therefore, because 'underlying it is an idea of science, of its aims and purposes, which is entirely different from mine' (Popper 1959, 80). This suggests that while Hesse's network model can be used against empiricism, and convincingly so, it cannot be used as readily to attack conventionalism. Any decision to do so must be based on other grounds. And a concern for the imprisoned scientist need not be one of them: Giddens (1976, 144) argues that asking 'How is it possible to get from one meaning-frame into another, if they are separate and self-contained universes?' sets up an insuperable problem only by posing it incorrectly in the first place. He regards theoretical systems not as discrete one *from* another but as mediated one *by* another.

In the second place – and, as I will eventually show, connected closely to this – once the specific privileges which empiricism accords to the observation language have been withdrawn, the network model still retains a more general privilege by separating scientific discourse from 'natural language'. Scientific discourse is supposed to grow by 'metaphorical extension of natural language' and to change partly as a result of the 'reinterpretation of some of the concepts of natural language itself', but to remain apart from and even superior to 'natural language' (Hesse 1974, 4–5). This will not do, and the hermeneutic school of German philosophy has established a programme which counters assertions like it, although in doing so I think it strengthens rather than weakens the network model.

In its early form 'hermeneutics', from the Greek *hermeneuein* (to interpret), was defined as the study of understanding or interpretation, and it became the epistemological foundation for the *Geisteswissenschaften*, the human sciences which had to reveal 'expressions of man's inner life' (Palmer 1969, 98). These were counterposed to the *Naturwissenschaften*, the natural sciences which had to study phenomena intrinsically 'silent about man' (Palmer 1969, 103). Much later in geography Kirk (1963) made a comparable distinction between the Behavioural Environment and the Phenomenal Environment, but the distinction made by the hermeneutic project was more than an ontological one; it was also an epistemological one, between *verstehen* (understanding) in the human sciences and *erklären* (explanation) in the natural sciences. These couplets stand in stark contrast to the methodological unity of *la certitude*, and they must assume particular force in a geography which is concerned in some sense with both the human and the natural worlds. This cannot be deflected by treating *verstehen* as 'fundamental to hypothesis testing' and then using a set of corroborative procedures to constitute *erklären* (Harvey 1969, 59). The early hermeneutic message is unequivocally 'two worlds, two geographies', and on a similar basis Tuan (1971, 182) has drawn attention to the difference between *existentialism*, which operates in a world of purposeful beings (and which from now on I shall call the life-world), and *environmentalism*, which operates in a world of objects.

Although Tuan's terminology is a loose one the distinction which it makes is one of both substance and method, and it cleaves geography in two. During the last decade or so it has encouraged human geographers to reconstruct perceptions of the human as well as the natural world and, more recently, to articulate clearly defined interpretative perspectives which allow them to enter more completely – to immerse themselves – in Tuan's world of intentionality and meaning. Their claim to do so has rarely been an explicitly hermeneutic one, and instead their models have been drawn from Weber and Husserl. They involve 'push[ing] one's own theories to the back of one's mind', a deliberate bracketing which is necessary because the human geographer, unlike his physical colleague, 'does not need theories of his own' all the time that 'he is concerned with the theories expressed in the actions of the individual being investigated' (Guelke 1974, 198).

Hermeneutics shares in this concern with the life-world and also refuses to impose a theoretical structure on it; but unlike the interpretative philosophies which have guided human geography up to now hermeneutics regards theories which are (partially) external to the life-world under investigation as *legitimate* prejudices which furnish the *necessary* conditions for any understanding: they must not be bracketed. Hermeneutics requires not so much the internalization of the experiences of the life-world demanded by Weber and Husserl but rather an interrogation of them. In this way *verstehen* comes to depend upon the *reciprocation* between two frames of reference rather than the *replacement* of one by another.

This is an avowedly creative process in which the interpreter is involved with but not enclosed by the life-world he is trying to understand, and since this is not in any sense 'a mysterious communion of souls' (Wolff 1975, 103) it presupposes that there is at least some common ground between the two, an arena in which the encounter can take place. As Habermas (1976, 134) has it, the structure of the life-world must 'previously be understood if the categories chosen [by the interpreter] are not to remain external to it'. But the pre-understanding is never total because the interpretative categories are never purely internal either: instead, the meaning ascribed to the one constantly mediates the meaning ascribed to the other through the 'hermeneutic circle'. Buttimer (1974, 24) sketches a similar strategy for geography which might make this clearer; she suggests that what human geography needs 'is an interplay of "inside" and "outside" views on situations: sensitivity to the native experience, and yet a grasp of the dynamics of situations from the external point of view'. Geography then becomes a *process*: 'native and non-native in dialogue developing a dramatization of the "geography" they see'.

An insistence on active dialogue recurs throughout the hermeneutic tradition, and this moves the human sciences in a dialectic between internal and external codifications, between what I referred to earlier as different theoretical systems. The human sciences depend on a mutually enriching and above all a reflexive relation between these various constructs, one which is missing from *le précis* and which requires a determined 'openness' about one's own preconceptions. I have suggested elsewhere that historical geography demands not only a critique of the constructs which survive from and were routinely drawn upon by historical actors but also and at the same time a critique of the constructs which the historical geographer himself brings to bear on the past (Gregory 1978). More generally, Gadamer (1975, 315) urges the need to 'be aware that there are always preconditions built into our social practice and organisations that enable us or hinder us in understanding one another'. Our incapacity to understand is thus 'rooted in what we are' (Taylor 1971, 51), and Gadamer describes hermeneutics as a discipline which 'guarantees truth' by 'making expressly conscious what separates us [from different codifications and theoretical systems to our own] as well as what brings us together'. Hermeneutic truth is intended to be a two-edged sword, therefore, cutting through illusions about life-worlds on one side and through illusions about ourselves on the other. As Tuan (1971, 181) observes, 'to know the world is to know one's self'.

The consequences of the early hermeneutic programme may appear unacceptable: it apparently dismembers an already loosely articulated discipline and substitutes a 'soft', romantic human geography for the 'hard', pragmatic science that we know today. There is nothing necessarily wrong with that, perhaps, but it need not happen: hermeneutics returns us very near to Hesse's network model. The network model is predicated on an almost identical process of progressive and reflexive investigation, but in the natural sciences. Putting the two together, the natural world is known through a dialectic between theory and observation, the human world through a dialectic between external and internal codifications. It is still the case that the human sciences obey what Giddens (1976) calls a *double* hermeneutic, in that unlike the natural sciences they are obliged to deal with a universe which is pre-interpreted, but it is nevertheless possible to use this first fusion to claim a universal application for the hermeneutic method. This has in fact occurred in two ways.

First, and most obviously given what has just been said, both the human sciences and the natural sciences can be regarded as intrinsically hermeneutic enterprises. Hermeneutics then becomes 'a universal aspect of philosophy, and not just the methodological basis of the so-called human sciences'; it seeks to understand 'the general relationship of man to the world', whether it be a world of purposeful beings or of inanimate objects (Gadamer 1975, 433). I will shortly suggest that this is inadequate – that is, that while (properly constituted) the human and the natural sciences are indeed hermeneutic, they have to be more than this – but as the argument has been presented so far it evidently overcomes the earlier threat of dualism and restores a methodological unity to geography's exploration of the human and natural worlds. This unity is, of course, vastly different from that envisaged in the positivist programme: whereas the Comtean model grounds all knowledge in a method which translates from the natural into the human sciences, the hermeneutic circle grounds all knowledge in a method which translates from the human into the natural sciences.

Secondly, the direction of this translation enables hermeneutics to assign itself a universal status which denies any form of scientific privilege. Its historical emergence from the human sciences means that it is concerned with the way in which man engages the world around him; it addresses *all* human experience and does not confine itself to 'science' and its modes of experience (Gadamer 1975, xviii). Ultimately, *hermeneutics does not regard itself as a privileged scientific 'method' at all but as the way in which man's appropriation of the world has to take place.* This mode of appropriation is a linguistic one: to Gadamer (1975, 431) language is 'a central point where "I" and the world meet'. This encounter is an ever-extending one, and although there are obvious affinities with Wittgenstein's 'the limits of my language mean the limits of my world' the horizon of Gadamer's world recedes (and the patterns within it change) through the iterative advance and intersection of the contours of language and experience. As Steiner (1975, 19) records, 'so far as language is mirror or counterstatement to the world, or most plausibly an interpenetration of the reflective with the creative along an "interface" of which we have no adequate formal model, it changes as rapidly and in as many ways as human experience itself'.

All this must seem a long way from ideology and science in geography, and yet it is absolutely central. If language has a speculative structure, if, as Gadamer (1975, 431) says, it is not 'the reflection of something given, but the coming into language of a totality of meaning', then discourse in geography necessarily shares in the construction of (and is in turn reconstructed by) that totality of meaning: in Lowenthal's (1961, 243) phrase, 'the universe of geographical discourse is not confined to geographers'. They – we – are not privileged observers of the world, if privilege is to be understood in any epistemological sense, but participants in it. To that extent, of course, geography has always and inevitably been 'relevant' to, and indeed in some measure responsible for, the structure of practical life. What hermeneutics invites (and what positivism specifically excludes) is an interrogation of the totality of meaning which 'relevance' presupposes, a continual examination of our mode of appropriation of the world.

Science, I will say, is obliged to be so self-critical if it is to distinguish itself from *ideology*, which I will represent as unexamined discourse. Any truly critical theory of knowledge 'can and must ask of itself what it asks of its object and subject-matter' (Valone 1976, 199), and this demands nothing less than an examination of language itself. (It also requires something more.)

According to Steiner (1975, 217–18) *'language is the main instrument of man's refusal to accept the world as it is.* Without that refusal, without the unceasing generation by the mind of "counter-worlds" – a generation which cannot be divorced from the grammar of counterfactual and optative forms – we would turn forever on the treadmill of the present'. This is where language reveals its creative as opposed to its purely reflective dimension: in its capacity to 'un-say' the world. Olsson (personal communication) argues:

the only way we can break out of the current world is to break explicitly and consciously with those internal relations that tie our words and worlds together into a coherent whole. In my vision, this means that we cannot really do any critical social science unless we begin to assault the very reasoning by which we have done our work in the past. So, perhaps, the real challenge we are now up against is to begin to write works which can do to the social sciences what Rimbaud and Mallarmé did to poetry, Joyce to the novel and Brecht to the theater.

Several geographers have risen to this challenge, some by turning to the philosophy of internal relations (Harvey 1973) and others by investigating the structures of alternative logical systems (Gale 1972; Olsson 1975). This is not the place to evaluate the success of their efforts, which are still in an early stage of development, and all I want to do here is to show why these responses have proved to be necessary: why, that is, positivism fails to provide the innovative vision of a critical science.

The heart of the matter is the distinction which positivism makes between *positive* theory, which claims to describe the world as it is, and *normative* theory, which claims to describe the world as it ought to be. Thus, for example, Chisholm (1971, 114) declared that 'however useful' human geography's emerging emphasis on positive theory 'may be in dealing with urgent and important problems of describing what "is" and forecasting what "will be", it will give us no insight into what "should be" ' and so 'there will be a continuing need for normative theory'. The reflective and creative dimensions of language are not skewered by the positive and normative distinction, however. Chisholm (1971, 130) suggests a technical reason for this when he recommends that 'normative theory must be static and not dynamic' because 'the moment that growth paths, and paths of change, are postulated, uncertainty finds the door wide open and comes bouncing in'. But the real reason, I think, and one which is revealed by Chisholm's rejection of uncertainty as 'a deadly enemy of normative thought', is an epistemological one: normative theory does *not* assault the categories which have become embedded in our discourse, and far from transcending them it merely re-orders them. It operates within the same unexamined discourse as positive theory and so necessarily replicates a structurally similar world. Smith (1977, 14) concedes that the distinction between positive and normative theory 'has become somewhat blurred in geography's relevance revolution'; but the point, surely, is that it is a contrived *ir*relevance. To generalize this, we can say that both positive and normative theory are articulated through a categorical paradigm, and that critical science is articulated through a dialectical paradigm. Labels like these inevitably conceal some differences beneath them, but they help us sort our intellectual baggage more easily.

The *categorical paradigm* encloses the world in a regular and unyielding linguistic mesh and conceives of change as a kaleidoscopic recombination of the fixed and precise categories which this immediately provides (Albrow 1974; Gregory 1976). The difficulty with this, as Harvey (1972, 327) recognizes, is that we are using 'stationary categories of thought to deal with a shifting universe', and it is through this that scientific discourse becomes if not privileged then at least recognizably different from 'natural languages [which] have evolved a looseness of structure and an inherent ambiguity which permits us to capture part of that shifting movement while using the same words'. The distinction is practical as well as linguistic, of course, and extends into the sphere of human action: as Olsson (1975, 12) says, 'the scientist tends to be certain about ambiguity, while the actor tends to be ambiguous about certainty'. The *dialectical paradigm*, by contrast, encloses the world in a linguistic mesh which changes coherently with the world and conceives of change as a fluid transformation of the transient and indeterminate relations which this provides (Ollman 1971). As Pareto observed, it was precisely this conception which Marx tried to capture by using words 'like bats', in which both birds and mice can be seen at the same time. Both Harvey and Olsson have grappled with these notions, and their work is obviously not without its practical consequences either: it allows for a genuinely socialist critique of everyday life.

There is rather more involved than this simple account makes out, as I will try to show in the next chapter, but it will serve as a first approximation. Here, I want to use these paradigmatic distinctions to confront *le précis* and *l'utile*. A double offensive can be mounted because critical science conjoins its rejection of these two precepts in its belief that science cannot prove itself by experimental means but only through concerted social action: 'the diagnosis that it offers to society, and its outline of future practice, can

C

prove themselves ultimately only in the free acknowledgement of those men who have experienced as real freedom an alteration of society deriving from this theory' (Wellmer 1971, 41). It is important to realize from this that the rejection of one precept interpenetrates the rejection of the other, and this makes it difficult to disentangle them for a discussion which fastens on each precept in turn. In what follows, therefore, some distortion is unavoidable.

The categorical paradigm is vitally necessary to the business of hypothesis-testing because it has the effect of holding the relationships between the categories constant (Harvey 1972, 327). It is easy enough to see how this operates *within* the formal models presented in the previous chapter: the theoretical terms are assumed to be fixed, the observational terms are assumed to be fixed, and corroboration is immediate and unambiguous. I have already shown, via the network model, that this assumption of stability is invariably violated and that its operational procedures are always disrupted by the dialectic between theory and observation. I now want to show that it is equally impossible to guarantee the stability of the relationships *beyond* the domain of the formal models.

Positivism typically attempts to do so through a *ceteris paribus* clause: using our earlier notation, this means that an hypothesis (*H*) is accepted (provisionally or otherwise) if the predicted event (*E'*) occurs *and provided other things are equal*. But suppose that *E'* does not occur: what then? This certainly does not falsify *H*, because other things may *not* have been equal: categories outside the model may have influenced those within it. In order to put this possibility to the test it is necessary to turn to some more general theory which incorporates both the external and the internal categories of the specific theory within its domain, and so specifies the effect of the one on the other. But the general theory will also be protected by a *ceteris paribus* clause, and to rely on its propositions is only to present the original problem in a more general form. In short, positivism has no alternative but to issue a disclaimer about the implications of any nominally unsuccessful test because it has no way of distinguishing the failure of the model from the failure of other things to be equal (Hollis and Nell 1975, 34; for a repetition of their argument in geographical terms, see Martin 1977).

Any distinction between positive and normative theory makes no difference to this result; both of them claim to show (in effect) 'what would happen, if . . .' *and neither of them can demonstrate that the provisions of this clause are fulfilled.* In saying that normative theory cannot be tested, therefore, Chisholm (1975, 126) must also recognize that positive theory cannot be tested either. The consequences – for positivism – are clearly disastrous. By insisting on empirical corroboration positivism necessarily makes its hypothesis irrefutable and, by its own definition, this prevents them from reaching the status of synthetic statements on which the progress of positive science depends. Positivism collapses into incoherence *on its own terms* and cannot realize its own project.

The consequences for other conceptions of science which choose to maintain what Popper calls an 'empirical basis' are perhaps less severe. Popper himself admits that his formulation of falsificationism is ambiguous, and he agrees that hypotheses and observations can always be harmonized with the aid of auxiliary propositions and clauses; intellectual honesty, he says, therefore demands some inter-subjectively acceptable (that is, negotiated) way of demarcating scientific *adjustments* from illegitimate *stratagems* (Popper 1959, 82–3; Lakatos 1970, 117). In other words, how do we determine (agree) when to protect a theory and when to suppress a theory? Popper's solution is to view empirical corroboration in the context of not one but a series of theories, and hence to deny that observation by itself can falsify an hypothesis; what is decisive is the emergence of a new theory in the series which offers excess information over its predecessors (Lakatos 1970, 120–1). This partially (but only partially) restores the theory–observation dialectic, and has been recognized in geography by, for example, Bird (1976, 156).

But the dialectic can be introduced in a much more prominent position, and in a role which Popper would undoubtedly turn down. The objections thus far have displayed the logical consequences for *le précis* of a commitment to the categorical paradigm; it is also possible (although Popper would dispute – or at least draw back from – this) to make a further objection on ontological grounds. To speak plainly, the critical sciences maintain that the object of their inquiry is a dialectical totality which *cannot* be reduced to a set of categorical constructs.

One of the earliest attempts to break with positivism and to come to terms with the implications which this has for geographical explanation was Gale's (1972; 1973) appeal to a logic of inexact classes as a means of developing a logic of locational decision-making. 'Fuzzy sets' are not dialectical relations, however, and Gale's is an inadequate foundation for a critical geography. Harvey (1972, 325) points out that 'there is no end to the number and variety of logico-mathematical devices which we can construct, and we are left with a marvelous weaponry (many of which are so heavy that scarcely anyone has the intellectual capacity to lift them) with no idea of how or why these weapons might be used. In short, we have the weapons but we have no idea what ammunition to use or where to direct them, because they are formal devices devoid of content.' In fact, however, we have rather *more* than an idea: Gale's project is 'oriented more towards solutions to practical problems and pragmatic questions' (Gale 1973, 262–3). Now, if it is the case, as he claims, that 'the questions are taken as the epistemologically primitive realisations of an inquiring mind from which the range of potential answers (the theories) and the strategies for providing answers (the methods) follow directly' (Moore and Gale 1973, 143), then it must also be the case that the questions which he is asking are specifically technical ones *and that they constitute a specifically technical conception of science*. It is in this sense more than any other that Gale's proposal 'compounds the error of the positivist model it is designed to replace' (Harvey 1972, 325).

This conclusion is a general one, and Habermas (1976, 154) has argued that the technical conception of science is implicit in Popper's treatment of negotiated demarcations. In the Popper scheme, 'the empirical validity of basic statements, and thereby the plausibility of law-like theories and empirical scientific theories as a whole, is related to the criteria for assessing the results of action which have been socially adopted in the necessarily intersubjective context of working groups'. Research programmes are therefore symptomatic of a much wider movement, namely 'a comprehensive process of socially institutionalised actions, through which social groups sustain their naturally precarious life'. And, according to Habermas, 'this occurs under experimental conditions which, as such, imitate the control of the results of action which is naturally built into systems of societal labour'. The imitative impulse is neither optional nor coincidental: what Habermas is saying here is that the labour process *has* to achieve technical control over its materials and hence *necessitates* a conception of science which can provide intersubjectively agreed (nominally successful) predictions about them. 'The technical recommendations for a rationalised choice of means under given ends cannot be derived from scientific theories merely at a later stage, and as if by chance. Instead, the latter provide, from the outset, information for rules of technical domination similar to the domination of matter as it is developed in the work process'. This, Habermas argues, is true of all sciences which retain the 'empirical basis'.

The distinction between 'policy-relevant' and 'policy-forming' contributions vanishes as a result of this analysis: as Lewis and Melville (1977, in press) put it, 'what the regional scientist passes on to "key decision makers" is not just a collection of ideas and measurements but a form of knowledge which is so structured by a particular concept of explanation that it may be used by them. By defining knowledge in this way the regional scientist is already formulating policy by reasserting the conventional boundaries within which a choice of policy must be made.' Theoretical structures are inextricably fused to social actions and, contrary to *l'utile*, the link between theory and practice is a necessary one and as such must override any idea that science can be politically neutral. Smith (1977, 16) may well want to claim that a major task of human geography is 'to develop intellectual structures to guide the evaluation of alternative spatial arrangements of human affairs in such a way that the arbitrary value content of any policy recommendations is minimised', but this is *precisely* what positivism offers; what makes it inadequate is that it conceals the *non*-arbitrary 'value content' and at the same time fails to uphold the validity of those contained within other conceptions of science.

This must be pressed further. Habermas shows that the technical conception of science is not self-sufficient because the process of negotiation on which it relies is an intrinsically hermeneutic one: only through an act of reciprocal communication can agreement be reached over its procedures. These two social imperatives, therefore, *production* (labour) and *interaction* (communication) constitute two distinct forms of knowledge by determining their respective objects of study and the criteria for making valid statements about them. Put this way round, science is always committed in some way, whatever its form, and the specific 'means' which it makes available cannot be divorced from the specific 'ends' which provide its own legitimation. Habermas calls these 'ends' 'knowledge-constitutive interests', and argues that they constitute two primary conceptions of science: the *empirical–analytic* and the *historical–hermeneutic*. These various relationships are summarized in Table 1.

If this thesis is correct, it means that both forms of knowledge flow from a necessity of interests to which we cannot properly object but with which 'we must *come to terms*' (Habermas 1972,

312). I am unwilling to put this any more definitely at the moment because the connections which Habermas makes between knowledge and human interests leave several important problems unresolved (see below, pp. 157–60). Furthermore, his conception of empirical–analytic science is not sufficiently distinguished from positivism, and realism would provide a better guarantee of technical control (Lobkowicz 1972). But if that is the case then, since Habermas rejects

Table 1

Form of knowledge	Social imperative	Knowledge–constitutive interest	Object of study	Criteria of validity
empirical–analytic	production	technical control	phenomena	successful predictions
historical–hermeneutic	interaction	mutual understanding	meanings	successful interpretations

the universal claims which the empirical–analytic and the historical–hermeneutic sciences make independently of one another and instead urges their fusion (better, their transcendence), then critical science will have to explore the conjunction we encountered earlier between the network model and the hermeneutic circle.

Critical science, then, is not a separate form of knowledge but a dialectical relation between the other two. It does, however, have its own knowledge-constitutive interest: it is committed to emancipation.

As a principle this is difficult to quarrel with, all the more so in this empty and unelaborated form, and on these terms few geographers would want to dissociate themselves from it. How many would be ready to retain a more explicit commitment in their 'professional' capacity remains an open question, though, and one which in itself reveals a widespread belief in the possibility, even the existence, of an interest-free geography. In any case, a conscious commitment to emancipation is hardly a guarantee of a critical science, inasmuch as this also requires a definite transformation of the existing structure of inquiry. Geography's humanist tradition, for example, has always been concerned to harness the discipline for the ultimate good of mankind. But if Gilbert and Steele's (1945) geography of human happiness differed in emphasis from Bunge's

(1973) geography of human survival, they were nevertheless united by a conception of science in which knowledge was inherently neutral and only became committed through its translation into social practice. Although they insisted that geography would be substantially incomplete unless that commitment were made – a somewhat different argument from those who feared that it would otherwise be starved of funds – their tacit acceptance of interest-free inquiry made their pleas for 'relevance' little more than demands for the extension of existing modes of inquiry into new areas of contemporary concern. In so far as these approaches, and others like them, were based on a positivist epistemology the solutions which they offered were technical ones, and no matter how successful they might have been technical solutions they remained.

As a result, many of the early attempts to construct a 'relevant' geography excluded any critical reflection on science itself, which was still thought to be constituted outside of the social practices to which it had to be 'applied', as a 'given' whose decisions had to be accepted irrespective of any prior claims or motives. The effect of this was severely to limit public participation in the political process, because it raised the geographer above any overtly sectional interest and removed him from any dialogue with those beneath him. From this unassailable vantage point the function of the new mandarin, as Harvey (1974, 102) calls him, was to provide 'objective' and, by implication, *unobjectionable* solutions, legitimized by the compelling force of science rather than the free consent of society. There are obvious problems in grounding political action in a consensus model of social practice, of course, but even so it is perfectly clear that if forms of knowledge are constituted by specific interests anyway then the privileged authority of the mandarin must be as spurious as it is precarious. Further, and connected to this, a strategic element of critical science's commitment to an emancipatory interest is the active and conscious involvement of all sections of society in its transformation, and in fact Habermas's recent attempts to develop a theory of communicative competence (see below, pp. 157–60) are intended to establish the very dialogue which a reliance on *l'utile* necessarily denies.

Human geography has been slow to recognize all of this, although it is over a decade since Pahl (1967) exposed the myth of a value-free geography and complained that while some geographers were held to be suspect because they were 'doctrinaire Marxists' none of their accusers would ever admit to being 'doctrinaire capitalists'.

But by the 1970s the debate had moved beyond these simple carica-
tures, and the connections between the structures of technical control and the structures of late capitalism had become, if not exactly
a commonplace, at least a familiar proposition within the subject.
This inevitably transformed the context of any further discussion,
because it meant that disputes over an epistemology which provided
for the universalization of technical control could no longer be
represented as narrowly academic skirmishes. The critique of a form
of knowledge became simultaneously the critique of a form of
society and, as Kasperson and Breitbart (1974, 58) warned, 'the
outcome will not be decided over sherry'.

Most of the critiques have been explicitly tied to an emancipatory
interest which, if it owes little to a reading of Habermas (and I do
not say that this is essential), is broadly conformable with his project. In *Social Justice and the City*, for example, Harvey (1973, 128)
suggested 'the replacement of manipulation and control with the
realization of human potential as the basic criterion for paradigm
acceptance'. He described the intellectual task as being

to identify real choices as they are immanent in an existing situation
and to devise ways of validating or invalidating these choices through
action. This intellectual task is not a task specific to a group of people
called 'intellectuals', for all individuals are capable of thought and all
individuals think about their situation. A social movement becomes an
academic movement and an academic movement becomes a social
movement when all the elements of the population recognise the need
to reconcile analysis and action [Harvey 1973, 149].

From a different perspective, existential rather than Marxist,
Buttimer (1974, 36) has similarly urged geographers to avoid 'a
reinforcement of the managerial and paternalistic bias which charac-
terised our efforts in the past' by regarding teaching and research
as 'inter-related experiences directed towards eliciting a consciousness among the people involved' which can eventually transcend
the 'various determinisms' of the life-world. The intention of both
these contributions is to realize what Marx once called 'man in the
whole wealth of his being, man richly and deeply mentally alive';
that is, to overcome the alienation of contemporary science and
contemporary society. A genuinely critical geography has to recognize the necessary connections between these two if it is to have any
practical efficacy, and a continued commitment to *l'utile* makes this
impossible.

What this suggests, in turn, is that critical propositions must be historically specific, unlike the laws of positive science which claim to be universal in space and time. This contrast can be sharpened by comparing two early conceptions of social practice, those of Marx and Comte. To oversimplify, Marx refused to accept that society's laws were eternal – in fact he dismissed the *Cours de philosophie positive* as 'wretched' – and replaced Comte's fear of their transgression with his own faith in their transcendence. Hence in 1865 Engels reminded Lange that 'to us so-called economic laws are not eternal laws of nature but historical laws which appear and disappear', so that the task of a critical science as he and Marx saw it was to reconstruct and to transform these historically specific moments and so secure a socialist society. Neither positivism nor Marxism have stood still in the intervening period, of course, and whether geographers want to align themselves with a Marxian conception of critical science is perhaps beside the point. But all the time that geography draws upon universal concepts in its analysis of historically specific societies and space-economies it must continue to perpetuate a permanent present.

The pressure to do so has proved to be a strong one, particularly since many of the concepts of human geography have been modelled on those of the natural sciences. Much early work in locational analysis and regional science was little more than social physics in a spatial context, and these comparatively simple applications and analogies have since been considerably extended: thus, for example, Wilson has provided advanced accounts of the mechanics of urban and regional systems (Wilson 1970; 1974); Isard and his colleagues have investigated the application of classical field-theoretical models of physical phenomena to space–time development models (Isard and Liossatos 1975; 1975a); and several workers have attempted to use catastrophe theory to characterize discontinuous change in space–time systems (Amson 1974; Mees 1975; Wilson 1976). But while it is plausible for physics and theoretical biology to claim a certain universality for their concepts, the consequences of the social sciences doing so are, at the very least, extremely problematic.

Human geography's reliance on neo-classical economics has had exactly the same effect: inasmuch as its doctrines are explicitly naturalist, the systems to which they refer are held in (or 'naturally' tend towards) a state of eternal equilibrium (Rowthorn 1974). As a result, as Massey (1973, 34) notes:

in most location theory, as in marginalist economics, the existence of numerous perfectly-competitive profit-maximisers, or, alternatively, of an oligopoly, is assumed as given – and consequently (which is the point) unalterable. The dynamic of the system as a whole is ignored. Thus, for instance, although both perfectly-competitive and monopolistic or oligopolistic situations are studied, they are analysed as separate situations, which might obtain perhaps in different places or in different sectors of the economy. The dynamic relationships between the two, and particularly the development of one from the conditions of the other, are ignored.

This is a necessary consequence of using concepts which refer to eternal processes of exchange rather than to specific relations of production, because they effectively freeze the historical transformations of social structures. This commitment to naturalism severely compromises the explanatory power and hence the practical efficacy of the social sciences.

Finally, contemporary human geography has frequently emphasized purely 'spatial' concepts, and has on occasion even seen itself as a 'spatial science'. When 'certain basic and primitive properties of space' are addressed in this way, 'the most basic of its elements and the simplest of its symmetries' (Cliff, Haggett, Bassett, Ord and Davies 1975, 1), the specific social structures which articulate the space-economy are replaced by universal geometric ones. The connections between social structures and space-economics are by no means clear-cut, of course, and to assume that there is any direct and immediate relationship, that the one reflects the other without either having any degree of autonomy, would be naive; but this is precisely the point. To concentrate on a universal spatial logic (Pahl 1975, 249–50) is to obscure the mediations which make human geography a distinctively *social* science. As Sack (1972, 75) observes, 'the usefulness of a geometry of location is not a matter of principle but of fact; for the physical sciences, and for the purely physical processes studied by geographers, the geometry of location is also the geometry of explanation'; but 'the facts concerning the usefulness of the geometric properties of this space for the explanation of human behaviour are not at all clear'.

When Comte made his three stages of knowledge correspond to three successive modes of action, he recognized that the supremacy of positive science had to depend on the continued existence of the industrial order to which it was committed and on the ability of its propositions to explain the forms assumed by earlier societies. The same has been true of human geography. We have to be careful about this, of course: it would be a mistake to assume that every human geographer who has drawn on these various universal concepts is an apologist for contemporary capitalism; but this does not alter their objective effects (Rowthorn 1974, 65). Neither does it change the way in which they have denied historical specificity and constituted knowledge as a cumulative system: even some historical geographers have regarded the past as a domain in which to test intrinsically timeless theories. Quite obviously, then, if the propositions of critical science are to provide knowledge about historically specific societies it is difficult to see how they can be cumulative in the sense envisaged by *le relative*.

In case these remarks are misunderstood, I should say that I am not trying to resuscitate old arguments about uniqueness: whether geography ought to be primarily an idiographic or a nomothetic science. These oppositions are, in any case, unhelpful. What I am saying is that a critical geography must see it as an important *political* task to resist the integration demanded by *le relative* and that it therefore cannot afford to model itself on the natural sciences. As Giddens (1976, 13) has remarked, 'a sort of yearning for the arrival of a social-scientific Newton remains common enough, even if today there are perhaps many more who are sceptical of such a possibility than still cherish such a hope. But those who still wait for a Newton are not only waiting for a train that won't arrive; they're in the wrong station altogether'. And far too many of them are geographers.

The integration of human and physical systems, I suggest, is not so much an epistemological problem as an ontological one. In these terms it is resolved every day that men appropriate their material universe in order to survive. The two worlds are necessarily connected by social practice, and there is nothing in this which requires them to be connected through a formal system of common properties and universal constructs. In reifying this sort of system, human geography must inevitably represent social structures and space-economies as parts of the fabric of nature, which man can regulate only within limits which he transcends at his peril. In short, Comte's final precept may not be an adequate way of tying together scientific knowledge: but it is a more than adequate way of tying up political action.

This discussion has only been a preliminary exercise, but several conclusions can be drawn from it which are quite fundamental to any account of ideology and science in human geography. I want to present them not as critical reactions to positivism, but as a set of affirmative statements which merit further investigation.

I think it is important to proceed in this way because to use the Comtean model and its variations as *constant* reference points is, in the end, a debilitating exercise. In the first place, I do not want to substitute one orthodoxy for another, but to explore some formulations based on different premises and leading to different conclusions from those to which traditional geography has made us accustomed. My own preferences must be clear enough by now, but the choice between one position and another ought not to be imposed by *fiat*. If we are to clarify our various concerns and to engage in any kind of debate with one another, it is necessary to open up these other perspectives without falling into what Popper calls reinforced dogmatism; otherwise epistemological discourse becomes as constrained and as compromised as it was before we started. My critique of positivism has, I hope, served its purpose, and now it ought to be possible to move forward in a more explicitly constructive manner. In the second place, therefore, some geographers *have* managed to sustain the break with positivism, one which I regard as crucial, and to move beyond a catalogue of its inadequacies to begin work on a series of practical reformulations. Their efforts are still in the early stages and, as will be obvious from the next section, much of the discussion continues to be couched in very general terms. But progress is rapid, and it is only proper to see these contributions in their own right, however tentative or half-formed they may be at present, and to locate them in a problematic which, although unequivocally opposed to positivism, is much more than a purely negative attack on it.

Summarily, then, I will argue that critical social science involves:

(1) *structural explanation*: a form of inquiry which locates explanatory structures outside the domain of immediate experience and which problematizes the relationship between theory and observation;

(2) *reflexive explanation*: a form of inquiry which mediates between different frames of reference and which problematizes their self-sufficiency;

(3) *committed explanation*: a form of inquiry which specifies its cognitive interest and which problematizes its legitimation.

All of these are in fact different dimensions of a single proposition, namely that the function of social science is to problematize what we conventionally regard as self-evident. The phrasing is Max Weber's, but the intention behind it is a much wider one than his sociology might indicate. In its simplest form, it suggests that what makes science necessary, what distinguishes it from commonsense understanding, is the existence of constraints and meanings on and in our actions which we typically take for granted. Their disclosure thus requires a conscious and deliberate effort on our part. The implication of my characterization of science, therefore, is not so much that ideology, by contrast, is *a*-structural, *ir*reflexive and *un*committed, but that it fails to *problematize* the issues set out above; it was in this sense that I referred to it earlier as an 'un-examined discourse'.

There is clearly much more to science than this: these are only minimal requirements. Even so, it is important to try to discover more about their bearing on human geography before moving on to elaborate a more sophisticated (and presumably a more exclusive) set of criteria. They clearly pose enough problems as they stand, and perhaps these ought to be signposted in advance so as to provide something of a counterpoint to the discussion which follows.

The most obvious difficulty is the extent to which they can be pulled together: structuralism (Chapter 3) is often considered in-compatible with phenomenology (Chapter 4) and both of them have been criticized for their political impotence (Chapter 5). That these complaints exist, of course, that there is a (necessary?) *tension* be-tween these different positions, does not immediately disqualify any attempt to bring them into *relation* one with another. Some of the objections are ill-founded in any case, and, more important, critical science must fail in its own terms if it draws back from the exami-nation of its own modes of explanation.

Two other difficulties which could be raised are much more parochial, and they revolve around the extent to which these three stipulations enable human geography to be constituted as a distinc-tive discipline and, following on from this, the ancillary role they thereby assign to physical geography. The first of these is easily disposed of: the fragmentation of inquiry between one subject and another is inimical to critical science, and it has no reason to respect existing disciplinary boundaries. If the discussion which follows appears to be insufficiently 'geographical', therefore, this is not only because geography is perhaps the last of the social sciences to take the claims of critical science seriously and so can furnish correspondingly fewer detailed examples; it is also – and much more – a necessary consequence of the practical commitments of the philosophy itself. This does not return us to the *hierarchical* array of sciences envisaged by Comte; instead, it treats knowledge in *relational* terms, so that what appears as, say, 'human geography' when grasped from one historically institutionalized viewpoint appears as, say, 'political economy' when grasped from another. A major task of critical science is to comprehend these internal relations and their transformations (see Ollman 1971). By extension, therefore, 'physical geography' has no need to worry about its ties with 'human geography'; not only are the physical and human worlds necessarily conjoined in social practice, they are also dialectically related within the totality of knowledge. To insist on the apodictic and enduring primacy of any *one* relation, however, as the physical/human geography debate sometimes appears to do, is to miss the point of a critical science. If 'there can be no second-class citizens among those who, for whatever reason, study the nature and operation of spatial process-response systems' (Chorley 1971, 107), then, one hopes, there can be no first-class ones either.

These brief remarks cannot settle the doubts by themselves, I realize, but it should now be possible to explore the three modes of explanation with a clearer understanding of what is at stake. Some may have already decided that this is too much, and that it is probably preferable and certainly safer to stay within the established problematic; in the concluding chapter I will try to summarize the consequences of their decision and contrast them with those of its alternative. But before we can do so we need to establish the critical model in more detail.

Part Two

Geography and Critical Science: Alternative Formulations

3 Structural explanation in geography

But if a savage or a moon-man came
And found a page, a furrowed runic field,
And curiously studied lines and frame:
How strange would be the world that they revealed.
A magic gallery of oddities.
He would see A and B as man and beast,
As moving tongues or arms or legs or eyes,
Now slow, now rushing, all constraint released,
Like prints of ravens' feet upon the snow.
He'd hop about with them, fly to and fro,
And see a thousand worlds of might-have-beens,
Hidden within the black and frozen symbols,
Beneath the ornate strokes, the thick and thin.

Hermann Hesse: *The Glass Bead Game*

Modern geography's first encounter with structural explanation took place in France and it was, in many ways, an uncomfortable and an inconclusive one. It began with a debate between Friedrich Ratzel and Émile Durkheim at the end of the nineteenth century, in the course of which Durkheim tried to assimilate human geography to his own conception of *morphologie sociale* and, through this, to include it in his construction of a realist social science. But although the subsequent development of the Vidalian school of *géographie humaine* clearly owed much to Durkheim, the 'appendicular' existence which he proposed for it (Febvre 1932, 34) prompted geography to retain its interest in society while firmly turning its back on social science (Claval and Nardy 1968, 117). Thus although the *tradition vidalienne* stood out from other European schools of geography by virtue of its 'vision of collective man' (Buttimer 1971, 1), it took little or no part in the elaborations and revisions of the original scheme made first through the exchange theory of Marcel Mauss and later by the linguistic structuralism of

Claude Lévi-Strauss. In fact, when structuralism was finally connected up to the discipline, some fifty years later, it was in a form which had few direct links with the Durkheimian tradition at all: this second encounter was derived from the operational structuralism of Jean Piaget and the structural-Marxism of Louis Althusser.

This lacuna poses problems of interpretation and presentation for the present account, and no solution can hope to be wholly satisfactory. In what follows I have tried to tackle it head on by giving a greater prominence to Mauss and Lévi-Strauss than my outline history might seem to warrant. This can be justified, I think, not only because their efforts are important in their own right, but also because they serve as something of a *primary* pivot around which to articulate the discussion. This needs clarification, since the story is an intricate one.

The first section takes Durkheim's view of human geography as its starting-point and documents what I regard as the failure of his project. Many reasons can be adduced for this, no doubt, but I suggest that it arose in part from Durkheim's inability to offer a genuinely reductive explanation of society, and it was this omission which Lévi-Strauss claimed to have put right. His programme is described in the second section, when contemporary geography's search for spatial order is related to the conception of structure which this provides. Durkheim also ran into difficulties, I argue, because the very idea of a general social science was unrealizable in the terms which he laid down for it. Various structuralist readings of Marx might claim to rectify this, but what is particularly significant about the intervention of Marxism, as the third section shows, is its opposition to linguistic structuralism and its condemnation of the importance which Lévi-Strauss attached to the symbolic and the synchronic.

While geographers were not especially slow to recognize the consequences which Lévi-Strauss's emphasis would have for their own studies – to Claval (1975, 273), for example, French geography was traditionally and predominantly 'a science of action, whereas structuralist interpretations are productive above all in the realm of representations, images and symbols' – not all of them were prepared to shoulder the full burden of the Marxian problematic, let alone the highly abstract formulations of Althusser, as a way of confronting the practical and the diachronic. Thus, Brookfield (1975, 194–8) wanted to know whether we needed 'full Marx', and concluded that we did not: 'it seems to me possible to adopt the framework of operational structuralism from Marx without at the same time accepting certain of his basic assumptions and the historical laws derived, through operational structuralism, from these assumptions'. This is more of an open question than Brookfield cared to admit, all the more so since it rests on an extremely contentious reading of Marx, but in asking it Brookfield obviously built on – or, better, drew back from – Harvey's (1973, 287–302) demonstration of several methodological convergences between Marx and Piaget. And again, what is particularly significant about the introduction of Piaget's genetic epistemology is that it also shares a number of assumptions and procedures with Lévi-Strauss's scheme (Gardner 1976). Through this explicitly liberal manoeuvre, therefore, Piaget came to represent something of a *secondary* pivot, between Lévi-Strauss and (via Marx) Althusser.

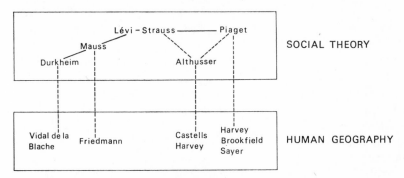

Figure 3   *Realism, structuralism and human geography*

These connections are summarized in Figure 3. They do not exhaust all the possible versions of structuralism which, as Poster (1976, 306) observes, has so far been more of a diffuse tendency than a neatly consistent doctrine. The picture is made all the more complicated by arguments over the domain of structuralism, and Chiari (1975, 162) notes that many are supposed, like Monsieur Jourdain who spoke prose without knowing it, to have been structuralists without being aware of it. The difficulty is in identifying an unambiguous touchstone, however, which would permit this sort of retrospective (and, for that matter, prospective) categorization. Structuralism is clearly not synonymous with structural explanation; but as soon as we try to sharpen its image, to focus the simple definition of the previous chapter, we are inevitably drawn into (perhaps appropriately) semantic debates about different conceptions of structure. I have not tried to resolve these terminological problems – many of them recall W. M. Davis's mindless arguments about the spelling of 'peneplain' – but instead to bring out the more substantial issues which lie behind some of them, and to concentrate on those which bear most directly on human geography.

Durkheim, realism and human geography

Durkheim was convinced that as long as the subject world, the realm of ideas, beliefs and representations, remained outside the object world its scientific investigation would be impossible. But this was no simple positivism, because he also scorned reductive explanations which tried to reduce 'psychic life to a mere efflorescence of physical life' (Durkheim 1893, 389) and instead announced his intention 'to bring the ideal, in its various forms, into the sphere of nature, with its distinctive attributes unimpaired' (Durkheim, in Wallwork 1972, 16).

This was to be achieved through what he regarded as the most fundamental of his *Rules of Sociological Method* (1895), the principle on which his entire programme was based: social phenomena were to be considered as 'things' ('*comme les choses*'), distributed along a continuum running from the spatial structure of society at one (morphological or structural) end to its rules, beliefs and emotions at the other (superstructural) end. His intention was thus to indicate that these various phenomena 'are not merely plastic creations of the will of the observer, but share the properties of physical objects in the sense that they exist independently of his observation of them. He cannot, therefore, discover their characteristics by *a priori* reasoning or by introspective examinations of his own consciousness' (Giddens 1972, 31). This did not mean a denial of any interior meaning they might have for individuals, only that their *scientific* investigation had to be an exterior one. This requirement bears the mark of Comte's much earlier insistence on *la certitude*, of course, but at the same time as Durkheim established this continuum he started to move along it and to transfer the locus of social explanation from the morphological pole to the superstructural one. It is this trajectory with which we are concerned.

In the *Division of Labour in Society* (1893) Durkheim had explained social differentiation in essentially morphological terms.

Traditional societies, he argued, displayed a *mechanical* solidarity: they were bound together by a belief system which allowed sub-groups to remove themselves without disrupting the material welfare of the community as a whole. Modern societies, by contrast, displayed an *organic* solidarity: they were bound together by a set of relationships which were so interdependent that sub-groups assigned to specific functions could not remove themselves without severe material dislocation to the community as a whole. Durkheim explained the transformation from one to the other by a 'law of gravitation of the social world'. This related the progressive division of labour to an increase in social interaction (which he called 'moral density'). But 'moral density cannot grow unless material density grows at the same time', so that 'the number and rapidity of the means of communication and transportation' were 'a visible and measurable symbol' of changes in moral density and could take their place in the general formula. Durkheim then appealed to Darwin and argued that as the contacts between traditional societies increased, so conflicts over the allocation of scarce resources between them became more likely, and they had to be resolved by a division of labour which transformed them into modern societies. 'Similiar occupations located at different points are as competitive as they are alike', he wrote, 'provided the difficulty of communication does not restrict the circle of their action' (Durkheim, in Giddens 1972, 150–4). In this simple model of a competitive space-economy, therefore, social differentiation was literally an agreement to differ, produced by changes in the spatial structure of society. The consensus might well be facilitated by some kind of cultural reinforcement mechanism, indeed it generally was, but its *explanation* remained a basically morphological one.

Geography had a part to play in this schema, and five years later Durkheim described it as one of the 'fragmentary sciences' which had to be drawn out of their isolation in order to contribute to a complete social science (Durkheim 1897–8, 520 and 1898–9, 556). He believed that it had to have a place in his projected mosaic because, so he claimed, Ratzel's attempt in *Anthropogeographie* (1882 and 1891) and *Politische Geographie* (1897) to change geography from an inventory into an explanatory science invited a reduction of geography to social morphology. The only difficulty he could see in this was that Ratzel had provided two conceptions of geography, and only one of them was admissible. 'Sometimes he clearly seems to propose as the object of political geography the settlement forms assumed by societies on the surface of the earth; and this, properly speaking, is social morphology. Sometimes he assigns it the goal of establishing the effects of physiographical features (rivers, mountains, seas, etc.) on the political development of these peoples' (Durkheim 1897–8, 531). Durkheim still regarded the first of these as the concern of any properly constituted social science, whereas the second accorded primacy to what he always insisted were entirely secondary environmental constraints. But if in Ratzel's eyes this was presumption rather than subsumption, if it emasculated geography and suspended society in the air, as he complained in the *Année Sociologique* of the following year, it was soon to be attenuated still further once Durkheim revised his earlier claims for morphological explanation. He continued to say that different forms of social action and expression depended on different spatial structures, but he was now disposed to allow these forms to become 'in their turn, original sources of influence' with 'an efficacity of their own': in short, 'partially autonomous realities which live their own life' (Durkheim, in Lukes 1973, 231–2).

This was developed most fully in *The Elementary Forms of the Religious Life* (1912), in which Durkheim tried to derive a general sociological theory of religion from what he believed to be its simplest form: totemism. His justification for this was that religion does not require communion with a god but 'a classification of all things known to men' as either *sacred* or *profane*. Totemism could therefore be the starting-point for a theory of religion even though it did not involve belief in an identifiable deity, because it designates particular natural objects as sacred, together with representations of them and members of the totemic clan who take some part in the rituals and ceremonials which surround them. All are in some measure sacred, and yet this is not a tripartite religion of totems, symbols and individuals 'but of an anonymous and impersonal force, found in each of these beings but not to be confused with any of them. None possess it entirely and all share in it. It is so wholly independent of the particular subjects in which it is embodied that it precedes them and survives them. Individuals die, generations pass away and are replaced by others; but this force always remains the same, real and vital.' It has 'no name, no history, abiding in the world and diffused in a countless multitude of things', 'the essence of so many different beings'. The totem itself 'is simply the material form in which the imagination represents this immaterial substance, this energy diffused through a variety of different things' (Durkheim, in Giddens 1972, 225–7, and in Aron 1967, 59).

This can be generalized to say that all social life consists in representations of some deeper reality which surfaces through them, and Durkheim maintained that this deeper reality is society itself. Participation in a religious ritual, therefore, is an unconscious affirmation of society, of the superiority of the community over the individual. Indeed, 'if we take away from man everything he derives from society all that remains is a creature reduced to sensation and more or less indistinguishable from the animal'. The commitment to society which is realized in ritual is thus an *essential* one which enables man to 'free himself of the domination of physical forces' and to find refuge 'in a force *sui generis*' (Durkheim, in Aron 1967, 101).

The claim which this evidently makes for the efficient causality of society (Aron's phrase) is, perhaps less obviously, also a claim for the existence of a scientific sociology, in so far as it specifically excludes two other potentially competing forms of explanation, *naturalism* and *psychologism*. First, representations in society cannot mirror patterns in nature because man has to distinguish himself from and even oppose himself to the physical world. By extension, therefore, Ratzel's claim for the efficient causality of the environment was also a claim for the existence of a scientific geography which could never be assimilated to Durkheim's conception of sociology. Secondly, representations in society cannot mirror categories given in the human mind because man has to enter into communion with others and become part of a totality whose *conscience collective* is different from and so irreducible to states of individual consciousness. By extension, therefore, psychology operated at a different level from sociology and could never be assimilated to it. Durkheim intended these two exclusions to mean that representations *in* society are really representations *of* society, and they allowed him to advertise what Hirst (1975, 81) describes as 'a vacant space awaiting its science'. Social phenomena had to be explained in terms of other social phenomena, and this was the proper concern of the new science of sociology, 'an object which belonged to it alone' (Durkheim, in Lukes 1973, 82).

The argument owes much to Comte in its claim for irreducible forms of social explanation, but it goes beyond the conception of causality which he took over from Hume in its simultaneous appeal to an 'essence', an 'immaterial substance', an 'underlying principle'.

This committed Durkheim to *some* kind of structural explanation, but not to any *particular* version of it. Even in *The Elementary Forms of the Religious Life* he was still searching for some correspondence between morphologies and representations (in the form structure ——> superstructure), while at the same time maintaining that 'the *conscience collective* is something other than a mere epiphenomenon of its morphological basis' (Durkheim, in Giddens 1971, 114). For all his confidence over the effect of society on the individual, then, he was now much less sure about the direction of causality within society itself.

To summarize, Durkheim's conception of sociology as an autonomous science of society required him:

(1)   to 'bring the ideal into the sphere of nature', in the methodological sense of adopting the same attitude towards both social and natural phenomena (which he believed would guarantee sociology's status as a science);

(2)   to 'remove nature from the sphere of the ideal', in the theoretical sense of explaining society in purely social terms (which he believed would guarantee sociology's autonomy).

This meant that he was unable to explain the emergence of society as something distinct from nature – to Durkheim 'society is ever-already present' (Hirst 1975, 124) – and although this may not be strictly necessary for the realization of a social science it is nevertheless desirable for one which makes such strong claims for an opposition between the social world and the natural world. It is, of course, this very difficulty which provides the starting-point for Lévi-Strauss's analysis of *The Elementary Structures of Kinship* (1949), a title which deliberately echoes Durkheim's. In trying to establish 'where nature ends and society begins', Lévi-Strauss arrives at an explanation of the *conscience collective* which, in contrast to Durkheim, suggests that the direction of causality within society can only be determined once this transition problem has been resolved. But whether this is true or not, and whether Lévi-Strauss's structuralism furnishes a satisfactory solution for it or an ideological gloss over it, the credibility of Durkheim's conception of social science remains threatened by its exclusive attention to the interactions between social morphology and social representation and its total silence about interactions with the natural environment. These have to have a place in any properly constituted social science because man is obliged to appropriate his material universe in order to survive and because he is himself changed through changing the world around him in a continual and reciprocal process which, following Giddens (1976, 100), I will call *structuration*. This preliminary definition of it will need to be elaborated later – the original phrasing is in Marx (1976, 283) – but even as it stands it is sufficient to indicate that Durkheim's hostility towards *géographie totale* was misplaced. In so far at Ratzel's proposals for it were 'neither environmentalism nor naive' (Dickinson 1969, 65), they were a reminder of Durkheim's inability to explain the structuration of specific societies and therefore a statement, however imperfect, of the limitations of his project.

These have since been pursued in various ways. Within geography itself, probably the most vigorous development from Durkheim's base-line was the school of *géographie humaine* inaugurated at the Sorbonne by Paul Vidal de la Blache. This interpreted regional man–land relationships in terms of distinctive *genres de vie*, and looked to livelihood to provide 'the label, the core around which a whole network of physical, social and psychological bonds evolved' (Buttimer 1971, 53). According to Vidal (1911, 194):

an established *genre de vie* implies a methodical and continuous – and therefore a powerful – action on nature. . . . No doubt man's actions have left a mark on his environment ever since he started to use tools; it could be said that his actions have been important since the dawn of civilisation. But the effects of organised and systematic customs are of a different order, as they dig their ruts deeper and deeper, imposing themselves with greater and greater force on successive generations, making their mark on the mind, and directing the march of progress itself.

The importance which this gave to the collective as opposed to the individual actions of men was characteristic of the Vidalian school as a whole and clearly owed much to Durkheim (Buttimer 1971, 51), but in its reluctance to stray further into sociological territory, presumably for fear of a counter-incursion, the school more or less confined itself to the way in which 'man has humanised the environment for his own ends' (Vidal 1921, 202) and avoided any examination of the internal articulation of *genres de vie*.

The historian Lucien Febvre correctly regarded this as a serious impoverishment of what the concept had to offer. Vidal had been right to regard man as a social actor rather than as a 'mythical abstraction', he said, but if geographers really were going to stop 'letting "Man" loose in "Nature" ' then they had to start looking at societies from something other than 'a "scenic" point of view' and to disclose the 'perpetual action and reaction' between them and their environments.

> To act on his environment, man does not place himself outside of it. He does not escape its hold at the precise moment when he attempts to exercise his own. And conversely the nature which acts on man, the nature which intervenes to modify the existence of human societies, is not a virgin nature, independent of all human contact; it is a nature already profoundly impregnated and modified by man [Febvre 1932, 161 and 363].

It can be seen from this, I think, that although Febvre had the greatest admiration for Vidal himself, and although he later said that *la géographie vidalienne* had played an important part in the formation of the *Annales* school of history, his interest in the structuration of specific societies set him some way apart from the mainstream of French geography. The Vidalian school continued to be preoccupied with a socialized nature: even Vidal had described geography as a natural rather than a social science, and many of the pupils who followed his teaching took this to mean that 'the nearer man is to the brute, the more geographical he is' (Febvre 1932, 364). The complexities of his social structures received little attention, and where they were examined at all the typology of *genres de vie* which resulted was a resolutely empiricist one with few concessions to any kind of structural explanation. This dismayed Febvre too, because although he was no structuralist he was undoubtedly influenced by several precursors of the structuralist school (Mann 1971). Durkheim was not the least of these, but again Febvre's debt to him and his relationship to the mainstream of French sociology was an equivocal one. The *Année Sociologique* had been 'one of my favourite intellectual mistresses', he confessed, but its unwillingness to allow any protracted intercourse between man and milieu was a major frustration.

Febvre was thus balanced on two tightropes which slowly moved apart: the gap between them is not only a measure of Febvre's personal discomfort but, more important still, it is symptomatic of the persistent lacuna which I identified at the beginning. The distance cannot now be overcome by somehow returning to the *status quo ante*, of course, since neither social theory nor geography stood still once they had disengaged from the dialogue. But the continued development of the Durkheimian tradition (and in particular its transformation into linguistic structuralism) has, I think, made its *rapprochement* with contemporary geography much less problematic than it once was. In part, certainly, this is no more than the result of new attitudes towards interdisciplinary initiatives, changes which have made the old intellectual imperialisms refreshingly redundant. But more immediate reasons can be advanced as well: the first, I suggest, is the dissociation of human geography from physical geography, a move which anticipates the divorce which Durkheim had wanted to make absolute, and the second is the rise of a spatial formalism within the regional sciences, a move which sacrifices the examination of specific content to the constitution of general structure. Both of these developments are commonplaces, and all that remains is for them to be brought into relation with the equivalent developments which took place within social theory.

Lévi-Strauss and linguistic structuralism

Lévi-Strauss owed much to Durkheim, but in his inaugural lecture at the Collège de France in 1960 he suggested that Durkheim's vision of social science was far too exclusive and its autonomy wrongly conceived:

> Too often since Durkheim – and even among some of those who believe themselves to be liberated from his doctrinal grip – sociology had seemed like the product of a raid hastily carried out at the expense of history, psychology, linguistics, economics, law and ethnography. To the booty of this pillage, sociology was content to add its own labels; whatever problem was submitted to it could be assured of receiving a prefabricated 'sociological' solution . . . we owe it in large part to Mauss and to Malinowski that we are no longer at that stage [Lévi-Strauss 1967, 13].

Their contributions are, therefore, perhaps the easiest *entrée* into his own project.

Mauss and exchange theory

Marcel Mauss was at once Durkheim's favourite nephew and his most distinguished pupil. After his uncle's death in 1917 he moved to the centre of the French sociological stage, and eight years later he completed his *Essay on the Gift: forms and functions of exchange in archaic societies.* When Lévi-Strauss came to read this, many years after it was first published, he recalled that it was 'like Malebranche hearing Descartes lecture, the heart throbbing, the head seething and the mind invaded by a certainty still indefensible but domineering, at having attended an event decisive in the evolution of science' (Lévi-Strauss, in Gardner 1976, 124). The verdict has stood the test of time, and Sahlins (1974, 149) describe's Mauss's essay as 'his own gift to the ages'.

What was so important for Lévi-Strauss was Mauss's clear demonstration of the need to go beyond the empirical to a more profound *structural* reality and to grasp not the particular and the conscious significance of social phenomena but their universal and *unconscious* significance.

Mauss's starting-point had been a very simple question: 'In primitive or archaic types of society what is the principle whereby the gift received has to be repaid?' (Mauss 1970, 1). Systems of primitive exchange were well known and had exercised generations of anthropologists, of course, but most of them had confined their efforts to the recovery of ethnographic detail: what Mauss proposed to provide was a general explanation for all of them. This was an extremely ambitious claim, and an example will reveal the enormity of the task. Probably the best known of all these systems was the so-called Kula Ring, which linked the Papuan islands of the Pacific Ocean through a clockwise transfer of long necklaces of red shell (*soulava*) and a counter-clockwise transfer of bracelets of white shell (*mwali*). 'Each of these articles, as it travels on its own direction on the closed circuit, meets on its way articles of the other class, and is constantly being exchanged for them' (Malinowski 1922, 81). But although the shells passed right round the ring, in continuous or episodic movement, punctuated by elaborate rituals and ceremonials, no individual ever did: 'the system operated principally by means of voyages in each direction outward from each point in the ring, rarely going further than two legs from the point of origin. Each voyage was reciprocated, so that the network consisted of an overlapping series of contact fields' (Brookfield and Hart 1971, 324). Commodity transfers were associated with these exchanges, but the shells themselves, the basis of the *kula*, were strictly non-utilitarian and their movements were clearly distinguished from the utilitarian transfers within the *gimwali* (Mauss 1970, 20).

Mauss's answer to riddles of this order of complexity was as simple as his question: the gift exemplified both material and moral life by sheltering a restless spirit whose ceaseless stirrings animated its progress around the circuit.

In this system of ideas one gives away what is in reality a part of one's nature and substance, while to receive something is to receive a part of someone's spiritual essence. To keep this thing is dangerous, not only because it is illicit to do so, but also because it comes morally, physically and spiritually from a person. Whatever it is, food, possessions, women, children or ritual, it retains a magical and religious hold over the recipient. The thing given is not inert. It is alive and often personified, and strives to bring to its original clan and homeland some equivalent to take its place [Mauss 1970, 10].

Subsequent criticisms and revisions of Mauss's solution need not detain us here (see below, pp. 99–102; also Ekeh 1974 for a summary); what is important is, first, the affinities with Durkheim, which are clear enough, and, secondly, the appeal to a logic of reciprocity. Mauss (1970, 77) made it perfectly plain that he was dealing 'with something more than a set of themes, more than institutional elements, more than institutions, more even than systems of institutions divisible into legal, economic and religious parts'; he was concerned 'with systems in their entirety', with seeing 'their essence, their operation and their living aspect' and catching 'the fleeting moment when the society and its members take emotional stock of themselves and their situation as regards others'. In effect, then, Mauss believed that a system of exchange works 'not on account of the norms which govern it considered simply as norms, but rather on account of the implicit structure of the system as a set of logical relations, *an implicit structure whose own inner logical necessity determines the norms, rather than conversely*' (Badcock 1975, 30; italics added). It was this glimpse into a structural domain – which went beyond anything which Durkheim had to offer – which Lévi-Strauss took over in displaced form and used to develop what he regarded as latent within Mauss's work, present but unexposed: namely, a structural *method*.

Cultural codes and structural methods

To Lévi-Strauss (1967, 11) the concept of a structural domain was less important than 'the very special way in which Mauss conceived of it: foliated as it were and made up of a multitude of distinct yet connected planes' which were 'manifested in experience – privileged instances which one can apprehend on the level of observation'.

The primary problem, as he saw it, was to reassemble these instances in such a way that their collation disclosed the domain from which they were derived. Linked to this was a secondary problem, that of providing a genuinely reductive explanation of social phenomena, one which would reduce the social (cultural) to the natural and at the same time maintain a clear distinction between them. He worked out a single solution for both of these puzzles in *The Elementary Structures of Kinship* (1949).

He began with two basic propositions. 'Wherever there are rules we know for certain that the cultural stage has been reached. Likewise, it is easy to recognise universality as the criterion of nature, for what is constant in man falls necessarily beyond the scope of customs, techniques and institutions whereby his groups are differentiated and contrasted.' It followed that any 'universal rule' would present 'without the slightest ambiguity, the two characteristics in which we recognise the conflicting features of two mutually exclusive orders' (Lévi-Strauss 1969, 8). He maintained that the most fundamental of these universal rules is the prohibition of incest.

This is given such prominence partly because it mediates between the natural and the social words in such a way that it forms the basis of society itself: 'by casting sisters and daughters out of the consanguineal group, so to speak, and by assigning them to husbands who belong to other groups, the prohibition of incest creates bonds of alliance between these biological groups, the first such bonds which one can call social' (Lévi-Strauss 1960, 32). This can easily be demonstrated by an example. The Kariera of Australia belong to one or other of four marriage classes (Banaka, Karimera, Burung and Palyeri) which are linked by marriage rules which oblige Banaka to marry Burung and Karimera to marry Palyeri. According to the accompanying descent rules, the children of a Banaka man and a Burung woman are Palyeri, while the children of a Burung man and a Banaka woman are Karimera. Similarly, the children of a Karimera man and a Palyeri woman are Burung, and the children of a Palyeri man and a Karimera woman are Banaka. This is summarized in Figure 4, which shows that the Kariera are divided into moieties which exchange sisters and daughters in each generation. That these transactions which bind the several factions together are regulated by an incest-taboo can be seen from the genealogical representation set out in Figure 5. For a Banaka man, for example, the transactions which would be *possible* but which are nevertheless proscribed are the marriage of his mother, his mother's sisters, or his sister's children (all Karimera); his daughters, or his father's sisters (all Palyeri); and his sisters, or his father's father's sisters (all Banaka). The only transactions which can be *realized* are with the Burung (Lévi-Strauss 1969, 156–60; Badcock 1975, 40–2). But the incest-taboo is also fundamental in a much deeper sense. When Lévi-Strauss (1967, 32) says that 'it *is* society' (my italics) he is referring not so much to the pattern of social exchange which it

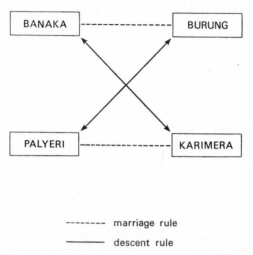

-------- marriage rule

———— descent rule

Figure 4   *Kariera marriage and descent rules (after Lévi-Strauss)*

establishes, but more particularly to the structure which is concealed behind it (and, by extension, behind all patterns of social exchange). This structure is contained in the method which he uses to uncover it:

(1) 'define the phenomenon under study as a relation between two or more terms, real or supposed;

(2) 'construct a table of possible permutations between these terms;

(3) 'take this table as the general object of analysis which, at this level only, can yield necessary connections, the empirical phenomenon considered at the beginning being only one possible combination among others, the complete system of which must be constructed beforehand' (Lévi-Strauss 1964, 16).

This is obviously the same sequence of operations which was used to analyse the kinship system of the Kariera, but it is also identical to the procedure established by Ferdinand de Saussure in his cele-brated *Course in General Linguistics,* which made a key distinction between *langue* (the collective institution we all share to make com-munication possible) and *parole* (the instance in which a particular communication is realized). The parallel is no accident: Lévi-Strauss (1967, 32) claims that through his work on kinship systems

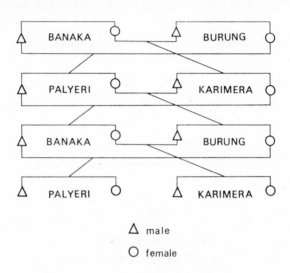

$\triangle$ male

$\bigcirc$ female

Figure 5   *Kariera kinship structure (after Lévi-Strauss)*

'what had been merely a huge and disordered scene became organised *in grammatical terms* involving a coercive charter for all conceivable ways of setting up and maintaining a reciprocating system' (my italics).

This can be pressed further, to say that Lévi-Strauss is disposed to regard all social phenomena as kinds of language or, better, as modes of communication. I think this is a preferable way of putting it because Lévi-Strauss repeatedly emphasizes the provisional nature of the analogy between language and society (Glucksmann 1974, 74–5). But, nevertheless, his intention is to explore the categories which man uses to apprehend, rationalize and explain the world around him, and he argues that although these categories are em-bodied *in* his various discourses and actions they derive *from* the structure of *l'esprit humain* itself. The linguistic analogy is clearly not far away and, in Goddard's (1975, 110) phrase, Lévi-Strauss's method constitutes an 'archaeology of consciousness' which simultaneously discloses a structural domain and offers a purely reductive explanation of Durkheim's *conscience collective*.

But what is even more distinctive about this solution is that Lévi-Strauss locates it within a Marxist problematic. He says that he accepts 'the undoubted primacy of infrastructures' and that he wants to contribute to the 'theory of superstructures, scarcely touched on by Marx' (Lévi-Strauss 1966, 130; see below, pp. 109–10). His contribution resides in his claim that there is always a 'conceptual scheme' which intervenes between infrastructure and superstructure, *and that it is confined within limits which are circumscribed by the unchanging human mind and its innate logic of classification.* But this means that Lévi-Strauss has an essentially categorical conception of historical change, in which 'through succeeding millennia man has only managed to repeat himself' (Lévi-Strauss 1955, 424). His actions are supposed to be determined by the models of intelligibility which he projects on to the world around him, and these in turn are 'the result of an endless play of combination and recombination' as he seeks 'to solve the same problems by manipulating the same fundamental elements' (Lévi-Strauss 1963, 8). In short, history becomes 'the process of playing through all the variants in an endgame of chess' (Hobsbawm 1973, 277), and the individual moves are of importance only in so far as they reveal the unchanging structure beneath them, that is, only to the extent that they represent 'privileged instances'.

What this means, of course, is that Lévi-Strauss cuts right through the dialectical conception of historical change and that, to continue the metaphor, he is not interested in the significance of individual moves for the outcome of particular games. Inasmuch as he accepts that 'interlocking states of thought do not succeed each other spontaneously and through the working of some inevitable causality' he is obviously not searching for some Hegelian Spirit to impart meaning and direction to history. But he also admits that there is nothing in the fundamental repertoire from which all these states of thought are derived which 'ordains that it should display its given resources at a given time or utilise them in a given direction' (Lévi-Strauss 1973, 474). The succession from one configuration to another is entirely beyond his provenance, and instead of trying to establish the historical relations between them Lévi-Strauss concentrates on their *logical* relations. This can be done in two ways: either various forms within the *same* mode of communication can be transformed

D

into one another (for example, kinship system A ——> kinship system A') or various forms within *different* modes of communication can be transformed into one another (for example, kinship system A ——> myth system *A*). It is through these logical operations which Piaget (1971, 98) calls 'transformation rules', that Lévi-Strauss claims to recover the general structure behind all social phenomena.

We must be quite clear about what this entails. Although Lévi-Strauss (1967, 17) conceives of (structural) anthropology as 'the bona-fide occupant of that domain of semiology which linguistics has not already claimed for its own', he is emphatically not advocating an analysis of the social construction of meaning. What he is doing instead, as Poster (1976, 319) puts it, is to displace the meaning-endowing subject altogether, by shifting the focus of intelligibility away from the subject to the *structure* and then by disclosing not the expression of meaning (the signified) but the pattern of its elements (the *signifiers*). Now it is, of course, possible to draw upon these structural methods without making Lévi-Strauss's heroic assumptions about the constitution of the domain which they uncover: the work of Friedmann, Godelier and Leach is surely a convincing enough testimony of this. But whether we agree that they reveal the innermost workings of the human mind or not, their minimal effect must be, as Lévi-Strauss (1966, 247) says himself, to dissolve the specificity of individual men and of their particular societies. *This is inescapable.*

It is now high time that we examined the implications of all this for contemporary geography. There are many avenues which could be taken, since the intersections between anthropology and geography have traditionally been as numerous as they have been important. But rather than suggest separate applications in, say, cultural and historical geography – perhaps the two most directly apposite sub-disciplines – it will be more helpful, I think, to pull these fragments together by considering the conceptions of spatial structure which might be provided by exchange theory and linguistic structuralism. In doing so, I want to begin to fashion a preliminary critique of their formulations, and so prepare the way for the examination of the encounter between structuralism and Marxism which follows.

Conceptions of spatial structure

Lévi-Strauss (1960, 52) once dismissed those who believe that structures lie at the empirical level of reality by saying that to them structure is like a jig-saw puzzle: if the pieces have been cut into seemingly arbitrary shapes, then there is no structure there at all. 'But if, as is sometimes done, the pieces are automatically cut in different shapes by a mechanical saw, the movements of which are regularly modified by a cam-shaft, the structure of the puzzle exists not at the empirical level: its key lies in the mathematical formula expressing the cams and their speed of rotation: something very different from the puzzle as it appears to the player.' It is in this sense, then, that the structuralist problematic conceives of structure. But from the discussion so far it is obviously possible to distinguish two levels of structural explanation in this, two senses in which 'structure' can be understood, since although both Mauss and Lévi-Strauss effectively reverse Durkheim's conclusions and suggest that, as it were, the first social categories are logical categories (Badcock 1975, 44), there is nevertheless an important distinction between them which can be made to correspond to two different levels of spatial structure.

When Lévi-Strauss (1966, 37) describes Mauss as a 'Moses leading his people to a promised land of which he would never contemplate the splendour' – in other words, towards but not into a general neurological science (Goddard 1975, 111) – he is himself preparing the way for a distinction between what I want to call a *collectivist* conception of spatial structure, which derives (in part) from exchange theory, and a *formalist* conception of spatial structure, which derives (in part) from linguistic structuralism. The brackets are necessary (a) because exchange theory contains an individualist as well as a collectivist school (Ekeh 1974), and since this resurrects the primacy of individual intentions and conscious decisions it is not open to structuralist displacements, and (b) because linguistic structuralism is certainly not the only source of spatial formalism within the regional sciences. These are only intended as provisional labels, then, which enable us to identify the major levels within the structuralist problematic – a context which must not be forgotten – much more easily than would otherwise be the case.

These levels are summarized in Figure 6, which brings them into line with Lévi-Strauss's jigsaw puzzle analogy and suggests a less

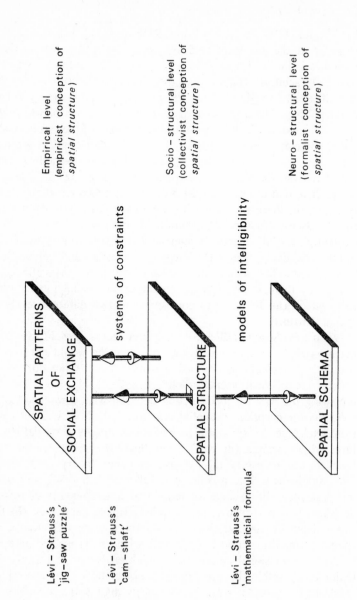

Empirical level
(empiricist conception of
spatial structure)

Socio – structural level
(collectivist conception of
spatial structure)

Neuro – structural level
(formalist conception of
spatial structure)

SPATIAL PATTERNS
OF
SOCIAL EXCHANGE

systems of constraints

SPATIAL STRUCTURE

models of intelligibility

SPATIAL SCHEMA

Lévi – Strauss's
'jig–saw puzzle'

Lévi – Strauss's
'cam–shaft'

Lévi – Strauss's
'mathematical formula'

Figure 6   *The structuralist problematic and levels of spatial structure*

ambiguous terminology of (spatial) pattern, structure and schema. In brief, and moving below the empiricist conception of spatial structure (here, *pattern*), the collectivist conception (*structure*) is concerned with the operation of deep-seated imperatives at a socio-structural level, and locates them there as a set of irreducible relations and transformations, whereas the formalist conception (*schema*) recognizes their presence at the societal level, but locates them at an even deeper, neuro-structural level.

With this in mind, we can proceed to clarify the systems of constraints and the models of intelligibility which I have suggested connect the two structural levels to the empirical.

First, there have been few theoretical elaborations of the collectivist conception of spatial structure within geography, and in so far as exchange theory has been drawn on at all it has been in largely informal and implicit terms which belong (in the main) to the individualist tradition. But it is still possible to isolate some contributions which, although couched in an individualist framework, nevertheless attempt to identify systems of constraints in a way which might be made consonant with a more collectivist tradition: or, at least, whose internal difficulties might be resolved if such a translation could be realized.

The most abstract of these is Friedmann's (1972) ambitious and programmatic project, which throws into relief the major forms of structural integration within an urbanized space-economy. This relies heavily on Peter Blau's *Exchange and Power in Social Life* (1964) for its construction of a two-dimensional matrix of exchange and power relations within urban systems (Figure 7). The typology which results is a suggestive one, but it is based on an unexplicated shift from Blau's original actor-centred model to what is clearly intended to be a structure-centred one. What is more, Blau argues that power relations are not simply linked to but are derived from exchange relations (Ekeh 1974, 182). Taken together, these considerations pose severe problems for those who, like Brookfield (1975, 121), see in Friedmann's work 'a link between Western regional development theory and some of the central issues of Marxist thought'. One of the most vital issues of all, in fact, is an elucidation of the reciprocal ties between actions and structures, since political practice must be predicated on a precise understanding of the mediations between the two if its interventions are to be effective. Furthermore, as Poulantzas (1973) is at pains to point out, Marxist thought regards power relations as class relations, and even

Exchange relations

		Reciprocal	Non-reciprocal
Power distribution	*Symmetrical*	*Fully integrated urban system:* moral authority dominates	*Competitive urban system* integrated on basis of limited liability: utilitarian power predominates
	Asymmetrical	*Active periphery of urban system* integrated on basis of protective dependency: utilitarian power predominates	*Passive periphery of urban system* integrated on basis of submissive dependency: coercive power dominates

Figure 7   *Structural integration within an urbanized space-economy (after Friedmann)*

if they are deeply imbricated in and elaborately embroidered by all levels of the social formation – which they are – they are ultimately relations of production and not of exchange. Now, whether Brookfield's extrapolation of Friedmann's project into Marxism is possible or not is beside the point if all we want to say is that a more rigorous foundation for the centre–periphery model, and certainly a more coherent formulation of its system of constraints, might be found within the collectivist tradition: it obviously sits very uneasily among the individualist assumptions of Blau's exchange theory anyway. But we must surely go further. Friedmann's project appears to anticipate our location of spatial structure at the sociostructural level, but its completion, I shall say, requires us to construct a conception of spatial structure in terms of what I have called the structuration of society. *And without taking the claims of Marxism seriously this is impossible.*

Secondly, there has been even less progress in the development of a formalist conception of spatial structure: so little, in fact, that some erstwhile pioneers appear to have given up altogether. Lowenthal (1961, 260) once held out the promise of a logical structure which unified the geography of the world, but his search for an adequate 'psychology of environment' left him with the feeling that he had 'conjured with stylistic configurations that may be too complex to characterise at all, let alone set in a comparative framework' (Lowenthal and Prince 1976, 130). But, as Tuan (1972, 321) notes, this sort of conclusion confuses the *existential* subject with the *epistemic* subject; while the first of these confronts what Lowenthal and Prince (1976, 130) correctly see as 'an ongoing flux, in which patterns are only fleetingly discerned and partially understood', the second discloses what Tuan (1972, 321) identifies as the 'cognitive nucleus which is common to all subjects at a certain level of abstraction'. This demands an explicitly structuralist approach, he maintains, which can reveal 'the human being and his response to the world *beyond the relativism of culture*' (Tuan 1972, 330; italics added).

It *is* possible, then, to discern a certain progression, a deepening, in our understanding of the relationships between nature and culture, starting from man's immediate and conscious speculations about it (Glacken 1967), going on to his symbolic and often unconscious images of it (Tuan 1974) and – this is the jump into the unknown – ending with the structural enclosures which underlie all these categories of thought.

If this sequence is to be completed geography will clearly have to overcome its data-bound conservatism: in Baker's (1977, 305) phrase, it will have to move beyond its traditionally narrow range of sources to consider 'drawings and dances, literature and landscape, music and memory, paintings and poems, rituals and rites'. What is important about all of these representations, according to Lévi-Strauss (1977, 77), is that 'they are permuted in different topological positions'; in other words, that 'what a given society "says" in terms of marriage relations is being "said" by another society in terms of village layout, and in terms of religious representations by a third'. The formalist conception of spatial structure can be established only through the collation of these privileged instances, scattered across space and through time: as their transformations are discovered, so the underlying scheme is reconstructed.

But this presents another, more profound, challenge, because geography will also have to overcome its theory-bound conservatism and examine the status and (above all) the practical efficacy of these structural enclosures. Following Lévi-Strauss, it would presumably be possible to argue that although spatial schemas are not self-evidently inscribed in spatial patterns, the one can be derived from the other by regarding (for example) the several point–process models which can replicate the patterns as a sub-set of the various models of intelligibility which man projects on to the world around him. They would then have their origin in the same invariant repertoire, the mind, and would therefore of necessity disclose configurations which recur time and time again. Gould (1974, 32) has admitted the possibility: 'what is surprising is not the uniqueness of patterns of spatial organisation, except in the most trivial sense of the word unique, but their extraordinary similarity. The constraints may vary from place to place, but upon their release familiar, consistent and non-surprising patterns evolve. There are, perhaps, deep structures of human behaviour underlying these repetitive patterns, if only we have knowledge enough and the eyes to see.' And as a matter of fact conventional pattern analysis often uses methods taken from information theory which interpret the realized configuration as but one instance of all the possible configurations which could be formed from its elements (Chapman 1977), a procedure which is formally equivalent to Lévi-Strauss's elementary methods of structural analysis.

Does this mean, then, that spatial structures are simply a product of a universal way of looking at the world – that their basic forms are no more than the limited combinations allowed by the mind's inner logic of classification? If this were the case then to say 'that there is more order in the world than appears at first sight is not discovered *till the order is looked for*' (Sigwart, in Chorley and Haggett 1967, 20; original italics) would *only* be to predict that the order will be discovered, and then *only* because the mind can structure its apprehension of the world in a limited number of ways. In short, spatial order would reside inside the mind and not inside the landscape, and the methods of conventional pattern analysis would 'reveal more about the language we are talking *in* than about the things we are talking *about*' (Olsson 1974, 53).

Yet this must be too restrictive. Lévi-Strauss makes it perfectly clear that these schemas are embodied in discourse and in action, so that while their constitution is certainly in part a product of man's contemplation of the world it is also a product of his participation in the world. Their models of intelligibility 'pass through the mind both ways', as Hillier and Leaman (1975, 9) put it, to reflect both cognitive and material appropriations, and in this sense their inner logic might enable us – or at least help us – to explain both the way in which the landscape is intelligible to us and the way in which we make it intelligible to others.

A project like this assumes, firstly, that we are able to relate our particular ('geographical') models of intelligibility to the total set from which they are drawn – in other words, that we are prepared to make a fundamental examination of discourse, both lay and technical – and, secondly, that we are able to accept that 'all systems by which mankind changes its relation to nature are also elaborated into systems of social signification' (Hillier and Leaman 1975, 9). Both of these ought to be unexceptional; but they demand a conjoint analysis of both sides of the equation and, I suggest, they condemn to failure any attempt to expose the structure of discourse or to isolate the semantics of spatial surfaces without regard to their material conditions. It is of the utmost importance to recognize, with Lefebvre (1974, 168), that *'cet espace a été produit avant d'être lu'* : that spatial structures are organized primarily to conduct historically specific modes of material appropriation and only secondarily, and via a series of historically specific displacements, to symbolize them.

Even so, what is missing from this account is any detailed statement of the way in which the two levels are articulated : what, precisely, is the relation between these spatial structures and their (?)corresponding schemas? Lévi-Strauss's allusions to Marxism are too fragmentary to provide much of an answer but, clearly, if we are to go any further we need to clarify the way in which these schemas are supposed to mediate between infrastructure and superstructure. And, again, *without taking the claims of Marxism seriously this is impossible.*

So far so good. But if the collectivist and the formalist conceptions of spatial structure are incomplete when considered separately, if we are unsure about how to consider them together, and if both these deficiencies require a careful evaluation of the claims of Marxism, there is still a long way to go. We must proceed with caution.

The structuralist–Marxist encounter

So far, I have suggested that an adequate conception of spatial structure has to be derived from an understanding of structuration, and in what follows I attempt to ground this claim more firmly. I do so by establishing in detail the concept of the mode of production on which, if my thesis is correct, both spatial structure and structuration must depend. But this immediate project is underwritten by a deeper one, because Marxism contains several competing concepts of the mode of production. These cannot be accepted or rejected on the basis of 'what Marx really meant': even if such a disclosure were possible, it would only resurrect the fideism which crucified debates through the 1950s and into the 1960s. What matters, on the contrary, and in both intellectual and political terms, is that Marxism should not be regarded as a pre-formed, pre-given science, but instead as one which requires constant and critical examination if it is to validate its self-conception and to fulfil its ultimate promise. But, as I must show, this begs the question. The encounter between structuralism and Marxism – whatever else we might make of it – has clearly provided, through Althusser's symptomatic reading of Marx in particular, a detailed construction of the concept of the mode of production; but, I shall say, *it has failed to provide a consistent location for science within (or without) the social formation*. The self-conception of Marxism in the present context is thus, at the very least, an ambiguous one. And this obviously reacts back on – and underscores the highly provisional nature of – our immediate project, to such an extent that this demands first, an initial statement of the Althusserian problematic, secondly, an outline construction of its major concepts, and thirdly, a critical summary of its consequences. Only then, I think, will it be possible to reopen our consideration of an adequate conception of spatial structure and to clarify its epistemological status.

Epistemological transformations

One of the key distinctions between Lévi-Strauss and Althusser is the way in which they treat the relation between logical and historical analysis. To oversimplify: where as Lévi-Strauss tracks back from the *apparent* movement of the social world to describe the structure behind it, Althusser tracks forward with the structure to describe the *real* movement of the social world in front of it. As it stands, of course, this is too blunt: in particular, it opposes the 'apparent' to the 'real' in terms which need a more careful phrasing.

As we have seen, then, Lévi-Strauss defines his object of study as a relation between two or more terms, and then constructs all the permutations which can be formed from them. He works like this, he says, because the mind is necessarily and immediately confronted by the manifest appearances of the social world, and its natural tendency is simply to operate with them and to produce variations of them. This may be one way of discovering an inner logic of classification, but all the time that the mind re-orders these basic categories it will continue to replicate a structurally similar world and in doing so it will encourage men to accept the basic conditions of their existence. Lévi-Strauss describes this categorical paradigm as *la pensée sauvage*, the savage mind, a mode of thought which he claims dominates primitive societies, the so-called 'cold' societies which 'are surrounded by the substance of history and try to remain impervious to it' (Lévi-Strauss, in Charbonnier 1969, 39). It follows from this that, to speak with Habermas, the savage mind can have no emancipatory interest, so that a major problem for a structural Marxism is to find some way of opposing it, some way of 're-estab-lish[ing] our connections outside ourselves, with the *very essence* of change' (Lévi-Strauss 1966, 256; italics added). It can only do so by refusing to work from the *apparent* movement of the social world.

Now, these simple oppositions were mirrored (and anticipated) in a displaced form in the work of Gaston Bachelard (1934), where they were conceived on two related levels:

(1) At the level of forms of knowledge Bachelard identified a constant dialectic between *reverie* and *science*. *Reverie* is 'the dreamlike character of everyday experience' and it represents 'the natural ["savage"] tendencies of the mind'. *Science* struggles against them and has to force man 'to make a break with the "spontaneous" interests of life' (Bhaskar 1975, 48).

(2) At the level of forms of existence Bachelard identified a constant dialectic between *phenomenon* and *relation*. In his view, every phenomenon is 'a tissue of relations' (Lecourt 1975, 39), but these interpenetrating webs are torn by everyday experience into discrete categories. These then become so deeply embedded in man's everyday discourse that they constitute an important obstacle to his scientific apprehension of the world.

Bachelard was not a Marxist, but his former pupil, Louis Althusser, used this system of concepts to examine Marx's writings, and in doing so he turned it into the equivalent system set out in Figure 8, which brings together these successive displacements and trans-formations.

Thus, to clarify, Althusser suggests that *ideology* fastens on the manifest appearances of the social world and reproduces them as categories in an unexamined discourse. This does not mean that it provides knowledge which is somehow 'false' – it does not have to

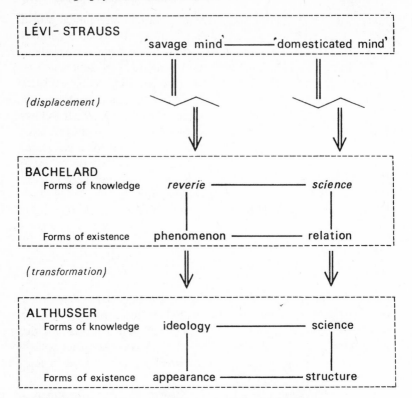

Figure 8   *Epistemological displacements and transformations*

do so at all – but that in failing to evaluate the categories with which it operates it necessarily fails to evaluate the world *on* which it operates. It legitimizes both a particular form of knowledge about the world and a particular form of existence in the world. *Science* therefore does not claim to provide knowledge which is somehow 'true' – to do so would be to accept the very concept of ideology which has just been denied – but instead to construct and reconstruct sets of theoretical relations which transform the structures of the social world: and again this has to be understood in both the cognitive sense of changing the way in which we know the world and the practical sense of changing the way in which we participate in it. Either way it will be a profoundly disturbing activity: 'science may invite us to give up ways of organizing our experience that seem to us so clearly correct, so forcefully guaranteed by what we know of the world by our ordinary experience of it, that we might prefer to think of our world as falsifying the theory rather than have the theory disrupt our relations with the world' (Mepham 1973, 116; see also Macintyre and Tribe 1975).

We might note here that this problematic detaches science from society in order to re-attach it as an autonomous process of critical construction.

Althusserian concepts of the mode of production

Marx constructed his theoretical relations through a critique of the then existing categories provided by classical German philosophy (Hegel), French socialism (Proudhon) and British political economy (Ricardo).

In the preface to his *Contribution to the Critique of Political Economy*, published in 1859, Marx recalled that 'the first work which I undertook for the solution of the doubts which assailed me was a critical review of the Hegelian philosophy of law'. Hegel had believed that man could only ever be directly aware of the manifest appearances of the social world, which he collectively called the State or Civil Society, but that each one of these phenomena was expressive of and reducible to exactly the same inner essence, which he called the Spirit or the Idea. The State was therefore 'the complete realisation of the Spirit in existence', 'the Divine Idea in so far as it is existent on earth', and 'the absolutely supreme phenomenal form of the Spirit'. The history of Civil Society was simply the progressive realization of the Idea – what Norman (1976, 7) describes as 'the unfolding of reason' – so that the historical transformation of society was contemporaneous with, joined to and determined by the underlying sequence of logical transformations.

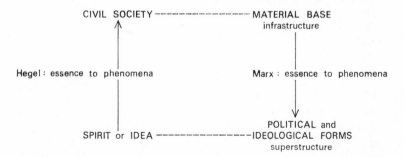

Figure 9 *Expressive totalities in Hegel and Marx*

But Marx was led to the opposite conclusion, namely 'that legal relations as well as forms of state are to be understood neither in themselves nor from the so-called general development of the human mind, but rather have their roots *in the material conditions of life*'. Essence and phenomena were transposed and idealism was replaced by materialism (Figure 9). As Marx put it:

in the social production of their life, men enter into definite relations that are indispensable and independent of their will, relations of production which correspond to a definite stage of development of their material productive forces. The sum total of these relations of production constitutes the economic structure of society, the real foundation, on which rises a legal and political superstructure and to which correspond definite forms of social consciousness. The mode of production of material life conditions the general process of social, political and intellectual life. It is not the consciousness of men that determines their existence but their social existence that determines their consciousness.

These classic statements have often been taken to mean that Marx's work can be collapsed into economic determinism: that, in the terms of our earlier argument, modes of communication are direct reflections of modes of appropriation and that the one can be immediately 'read off' from the other. And, certainly, if all Marx had done was to 'stand Hegel on his head' then the superstructure would have to be expressive of and reducible to the economic structure in the way that the State was supposed to be expressive of and reducible to the Spirit. But Althusser maintains that Marx did not regard the relationship between superstructure and economic structure as direct and immediate and that, in contrast to Hegel's 'expressive' totality, he saw these relationships as indirect and mediated through the various levels of a 'foliated' totality. According to Althusser, a symptomatic reading of Marx discloses three possible levels: the economic, the political and the ideological (Figure 10).

The economic level (or, strictly, the mode of production) is concerned with the way in which the *labour power* of the workforce uses the *means of production* to transform an *object of labour*. These three elements are articulated by a specific mode of appropriation of nature (or, in a more limited sense than we have used it up to now, a specific *mode of material appropriation*), and at the end of the labour process 'labour has become bound up in its object:

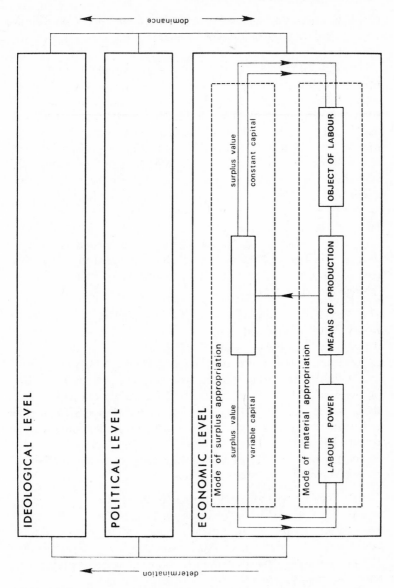

Figure 10 *Althusser's conception of the mode of production*

labour has been objectified, the object has been worked on' (Marx 1976, 287). Two propositions flow from this:

(1)   The false polarization between either the naturalization of man or the humanization of nature which we encountered in the debate between Durkheim and Ratzel is replaced by the dialectical concept of *structuration*, in which there is a constant and reciprocal interaction between both man and nature (Burgess 1976, 3); as Marx (1976, 283) put it, 'labour is, first of all, a process between man and nature, a process by which man, through his own actions, mediates, regulates and controls the metabolism between himself and nature'. 'Through this movement', Marx continued, 'he acts upon external nature and changes it, and in this way he simultaneously changes his own nature.'

(2)   The value of the end-product of the *labour process* is measured in terms of the labour which is embodied in it, and this then has to be transformed from its commodity to its money form through the *realization process*.

The second proposition forms the starting-point for Marx's labour theory of value, and although I do not want to present it in any detail – and for reasons which I will shortly indicate; in any case, excellent introductions are to be found in Dobb (1973) and Howard and King (1975) – some preliminary comments are in order. For purposes of clarity the discussion will be mainly concerned with the capitalist mode of production; Hindess and Hirst (1975) provide a detailed account of pre-capitalist modes of production which is broadly consistent with an Althusserian problematic.

The value which is embodied in the end-product of the labour process is made up of three components:

(a)   The first component is returned to the workforce in order to ensure the reproduction of labour power. The labour which the worker has to put in to produce commodities whose value is equivalent to this first return flow is called necessary labour, so that the value of labour power is itself the value of necessary labour or, formally, *variable capital*.

(b)   The second component is returned to the labour process in order to renew the means of production and the objects of labour which were used up in the preceding cycle. This second return flow is governed by the initial value of these elements, since 'the maximum loss of value [they] can [have] suffer[ed] in the process is plainly limited by the amount of the original value with which they entered it, or in other words, by the labour-time required to produce them'. They can therefore 'never add more value to the product than they themselves possess' (Marx 1976, 314), and since this original value cannot be changed during the subsequent labour process, formally, both the means of production and the objects of labour represent *constant capital.*

(c) The third component is returned to extend rather than to maintain investment in the labour process, and so promotes what Marx called extended reproduction. It is equivalent to the labour which the worker puts in over and above necessary labour: surplus labour or, formally, *surplus value.*

Surplus value is appropriated through a specific *mode of surplus appropriation.* According to Althusser, Marx's thesis was that in pre-capitalist modes of production the worker was not separated from the means of production, which meant that the surplus had to be extracted from him by *political* or *ideological* mechanisms (usually through the institutions of the State or the Church), whereas in the capitalist mode of production the worker is separated from the means of production and so has only his labour power with which to make a living, which means that the surplus can be extracted from him through the *economic* mechanism built around the wage-labour market. Two further propositions flow from this:

(3) Any society must depend on the two modes of appropriation which articulate the economic level, the 'real foundation' as Marx called them, and for this reason the full structure of economic, political and ideological levels is often identified in terms of its *mode of production* alone. But this is more than just a convenient shorthand because:

(4) Although different levels *dominate* different modes of production in order to appropriate the surplus, which level will dominate which mode of production is *determined* by the economic level.

In the capitalist mode of production, therefore, the economic level is both dominant and determinant, whereas in other modes of production other levels occupy the dominant position, but still as an effect of the conditions of existence of the economic level (Hindess and Hirst 1975, 14; see also Althusser 1969; Althusser and Balibar 1970).

This evidently does not mean that these other levels can be reduced to the economic level, and neither does it mean that they are so independent of one another that analysis is condemned to 'following the empirically diverse concatenations at different times of essentially separate and only contingently related factors' (Garaudy, in Callinicos 1976, 41). Instead, Althusser's construction of the concept of the mode of production provides a determinate mode of articulation both within and between the levels.

This distinguishes it from Harvey's (1973) construction, and a comparison between them will serve to sharpen the outlines of the preceding sketch. Harvey (1973, 199) claims that the mode of production is articulated through three 'modes of economic integration' (reciprocity, redistribution and market exchange), which were first formalized by Polanyi's substantivist school of economic anthropology. Althusser would have to resist this construction on three interrelated grounds:

(a) Polanyi's scheme is an *empiricist* one, and he regarded reciprocity, redistribution and market exchange as forms to be identified empirically and not as structures to be constituted theoretically (Polanyi 1968, 149). In complaining that the variety of forms in the economic, political and ideological levels prevents 'a unique characterisation' of a particular mode of production, therefore, Harvey fails to recognize that it is not the forms which identify the mode of production but the concept of the mode of production which identifies the forms (Harvey 1973, 205).

(b) Polanyi's scheme is a *categorical* one, in which historical change becomes a kaleidoscopic repatterning of these discrete forms into different combinations which represent different societies, and not a dialectical one in which one structure is transformed into another (Polanyi 1968, 156).

(c) Polanyi's scheme is an *idealist* one, in which the projected sequence of dominant forms is determined by and reducible to the progressive realization of a 'market principle' (Dupré and Rey 1969), and not a materialist one in which man's ideas are shaped by his social existence; the same is true of Harvey's (1973, 207) suggestion that 'the degree of market penetration into human activity' affords a useful characterization of social structure.

I have examined these issues in more detail elsewhere (Gregory and Smith 1977), and we can simply conclude that the conjunction of Polanyi and Marx must reduce any explanation to incoherence. But what is particularly important in the present context is that Polanyi never included modes of material appropriation in his substantivist problematic (Hindess and Hirst 1975, 26); since the omission cannot be made good by connecting his scheme to a Marxist one – Harvey's attempt to do so is little more than the kind of transliteration which Althusser (1969, 203) so roundly condemns – it must follow that concepts of reciprocity, redistribution and market exchange *cannot* be brought to bear on the concept of structuration and hence, for all their manifestly 'spatial' connotations, *cannot* constitute an adequate conception of spatial structure. This is true, I think, whether Althusser's problematic is accepted or not, and it is to this that we must now turn.

Critical comments

Althusser has not been without his critics, and I want to indicate two qualifications here, both of which are in themselves Marxian.

The first is that there have been many attempts to dissociate Marx from structuralism altogether, and to object to the elements of the linguistic model which are supposed to be embedded in Althusser's work (Glucksmann 1972). In the same way that Lévi-Strauss provides an account of *la pensée sauvage* rather than of *la pensée des sauvages* (Godelier 1973, 65), so Althusser's construction has been dismissed as a mere formalism, empty of all content: a 'complete mythology' of an 'abstract world of structures and levels' (Laclau 1975, 104). Moreover, in so far as his failure to present structures in historical terms – 'Spinoza once again at rest', as Hindess and Hirst (1975, 316) describe it – implies an acceptance of Lévi-Strauss's view of the synchronicity of structures, then Althusser is held to negate the role of continuous and conscious human action in transforming them.

These misgivings are, in principle, mistaken. Many of them have arisen as a result of some obvious infelicities in Althusser's terminology, which he has admitted is 'too close in many respects to the "structuralist" terminology not to give rise to ambiguity' (Althusser and Balibar 1970, 7). To repeat: Althusser does not divorce logical analysis from historical analysis, like Lévi-Strauss, and neither does he conflate them, like Hegel. Instead, he regards the one as a prerequisite for the other so that, to paraphrase Giddens (1976, 120), a structure can be described 'out of time' but its structuration cannot.

But the real difficulty is in fact Althusser's inconsistent realization of his project: he follows different protocols in his examination of science and in his examination of ideology. As Hindess (1977, 206-7) has shown, Althusser examines *scientific* formulations in relation to the *concepts* of their discourse (for example, logical dependence and coherence) but *ideological* formulations as an index of the *process of production* of their discourse. This is clearly teleological, inasmuch as 'the known conclusion determines how the discourse is to be approached in the first place' (Hindess 1977, 208). And this is no accident; far from being a contingent outcome of Althusser's problematic, *it is built into its very structure*. As we have already seen, Althusser's concept of the mode of production removes science from the social formation but implicates ideology in it. Hence, the autonomy conferred upon science means that it has to be examined 'internally', in terms of the concepts of its discourse, while the practical foundation of ideology means that it has to be examined 'externally', in terms of the process of production of its discourse.

Hindess (1977, 224) is surely correct to insist that *all* discourse ought to be examined in the same way, but he maintains that this has to be (primarily) in internal terms which 'cannot proceed by reference to any alleged process of production of that discourse'. This solution is, I think, one-sided, and it fails to clarify the status (location) of science. If we are to understand the constitution of both science and ideology it must be through an examination of discourse in both internal and external terms, and while the two must certainly not be confused it is important to spell out the way in which the double valency contained in the examination of discourse is to be articulated.

My second qualification, on a different but no less fundamental plane, is that the labour theory of value is also fraught with difficulties. I cannot discuss all of them here, but in the present context it is important to note Hodgshon's (1976, 9) claim that the concept of embodied labour is essentially an idealist one: that 'it is a pure thought-construct, rather than a concept which relates to any identifiable process or relation in any mode of production'. This is not, definitely not, the standard empiricist objection to embodied labour as a 'metaphysical philosopher's stone', because what is so important about Hodgshon's argument is that it is conducted (at least in part) from within Althusser's own problematic and hence that it rebounds on Althusser's construction of the concept of the mode of production.

Neither of these problems can be explored in the present text, but Hindess and Hirst (1977) have suggested that the way forward is through a reconsideration of the basic concepts of Marxism and, in particular, through a switch from the mode of production to the *social formation* as the touchstone of Marxian analysis. This immediately limits the prior claims which can be made for theoretical conceptions.

At most the concept of a determinate social formation specifies the structure of an 'economy' (forms of production and distribution, forms of trade, conditions of reproduction of these forms), forms of state and politics and forms of culture and ideology and their relation to that economy, economic classes and their relations, and the conditions for a transformation of certain of these forms. It does not designate a social totality with its necessary effectiveness given in its concept, nor, as a consequence, particular 'states' of the action of such a totality or of its resolution into some other form of totality. Neither the persistence nor the supersession of the economic, political or cultural/ideological forms can be deduced from the concept of the social formation in which they appear. In particular, the social formation cannot be resolved into the classical Marxist formula of economic base and its political–legal and ideological–cultural superstructures. Legal and political apparatuses and cultural or ideological forms provide the forms in which the conditions of existence of determinate relations of production are secured, but they are not reducible to their effects and they are not organised into definite structural levels which merely reflect the structure of an underlying economic base. This means that political forces and ideological forms cannot be reduced to the expression of 'interests' determined at the level of economic class relations [Hindess and Hirst 1977, 57].

This, incidentally, and as Stedman Jones (1977) pointed out, is the very trap which captures Harvey's conception of capitalism: as a matter of historical record it is clearly *not* the case that 'in a conflict between the evolution of the economic basis of society and elements in the superstructure, it is the latter which have to give way, adapt, or be eliminated' (Harvey 1973, 292; see also Brookfield 1975, 196).

Many points in this dense passage deserve comment, but as far as our immediate project is concerned its most significant result is its rejection of the concepts of base and superstructure *and* of structural causality. The first of these is straightforward, and makes Lévi-Strauss's original formulation of a schema interposed between base and superstructure even more untenable. (Which is not to say that the constitution of schemas is un*important*.) The second distances Marxism still further from the trappings of structuralism, and suggests that conceptions of spatial structure ought to be located in the concept of a determinate social formation; but in doing so, as I shall show, Hindess and Hirst allow for the analysis of spatial structures in a much more direct way than Althusser.

Modes of production, social formations and spatial structure

The most fully developed translation of Althusser's problematic into spatial terms is without doubt Manuel Castells's *sociologie urbaine* (Castells 1972; 1976; 1977; see also Short 1974; Pickvance (ed.) 1976). Castells develops a spatial characterization of the three levels of the mode of production as follows:

(1) *Economic level*
(a) *Production*: 'the ensemble of spatial realizations derived from the social process of reproducing the means of production and the objects of labour' (Castells 1977, 129);
(b) *Consumption*: 'the ensemble of spatial realizations derived from the social process of reproducing labour power' (Castells 1977, 130);
(c) *Exchange*: the ensemble of spatial realizations of transfers within and between production and circulation and 'which can be understood not in itself but in terms of the elements it connects' (Castells 1977, 130).

(2) *Political level*
The *institutional* organization of space, which Castells (1977, 208) takes to be bipolar: 'the state apparatus not only exercises class domination but also strives, as far as possible, to regulate the crises of the system in order to preserve it'.

(3) *Ideological level*
The *symbolic* organization of space, which Castells (1977, 127) identifies as 'a network of signs, whose signifiers are made up of spatial forms and whose signifieds are ideological contents, the efficacity of which must be construed from their effects on the social structure as a whole'.

At the heart of this system of concepts are the assumptions that spatial theories *express* social theories and that spatial structures *realize* social structures. To clarify: Althusser (in Althusser and Balibar 1970, 99) maintains that because Marx did not follow Hegel in reducing history to a single essence which unfolds through time it is not possible 'to think of the process of development of the different levels of the whole *in the same historical time*'. Instead, it is necessary to assign to each level a peculiar time or *temporality*. Thus, for example, the economic level of the capitalist mode of production contains 'different rhythms which punctuate the different operations of production, circulation and distribution', and the concepts of these different temporalities have to be constructed 'out of the concepts of these different operations'. But in so far as history is not only an interlacing of times but of spaces as well (Vilar 1973, 188), then it must also be necessary to construct the concept of the spaces of the different levels *in exactly the same way*. Indeed, Lipietz (1975, 417) terms them 'spatialities'.

It may well be, of course, that the disjunction between temporality and spatiality is incorrect, and that analysis ought to be directed towards 'an historically defined space–time, a space constructed, worked, practised by social relations' and organized 'into specific, articulated units according to the arrangements and rhythms of the means of production' (Castells 1977, 442–4; see also Thrift 1977, 68–9). But it will still be the case that, as Lefebvre (1974, 345) put it, '*le concept de l'espace n'est pas dans l'espace*'.

A recommendation like this is clearly capable of several inflections, but although Castells, like Althusser, is keen to dissociate himself from 'accusations' of structuralism, the conception of spatial structure which he provides, particularly in his early formulations, cloaked in the language of 'expression' and 'realization', is still inculpated in the general failings of the structuralist problematic. These difficulties even reappear in Santos's designation of the concept of social formation. While, with Peet, he wonders whether 'its omission has not been one of the factors retarding theorization in geography and preventing it from finding at the same time more concrete and epistemologically more coherent bases' (Santos and Peet 1977, 2), he nevertheless characterizes it in almost explicitly structuralist terms: 'the mode of production amounts to only a possibility to be realized, and the socio-economic [read 'social'] formation alone would be the *realized possibility*' (Santos 1977, 5; original italics). To speak with Hindess and Hirst, spatial theories *are* necessarily social theories, but spatial *structures* are not *given* in social *concepts*, whether these are embedded in conceptions of modes of production or social formations. This in no way denies the existence of a hierarchy of concepts, but simply insists that there can be no immediate 'correspondence' between structures and their concepts (Hirst 1977).

Hindess and Hirst (1977, 63–72) suggest that the concepts of 'possession of' and 'separation from' certain of the means of production are central to the analysis of economic classes within determinate social formations. They define the means of production as 'all the conditions necessary to the operation of a particular labour process which are combined in the units of production in which that process takes place. If any of these conditions is exclusively possessed by a definite category of agents, and the agents who direct or operate the labour process are separated from them, then such relations provide the basis for class relations'. What this means – and this is crucial – is that '*relations between the "unit of production" or "enterprise" and the systems of circulation or distribution of the conditions of production must be analysed if class relations are to be rigorously determined*. This analysis must be conducted for the concepts of the social relations of specific social formations' (Hindess and Hirst 1977, 65; original italics). Put this way round, the analysis of spatial structure is not derivative of and secondary to the analysis of social structure, as the structuralist problematic would suggest: rather, each requires the other. Spatial structure is not, therefore, merely the arena within which class conflicts express themselves (Scott 1976, 104) but also the domain within which – and, in part, through which – class relations are constituted, and its concepts must have a place in the construction of the concepts of determinate social formations.

Now, it is clearly important to transcend geography's 'fetishism of areas' and to 'destroy the myth that areas, *qua* areas, can interact' (Carney, Hudson, Ive and Lewis 1976, 13), but it should now be equally obvious that this must mean more than a simple demonstration that the spatial lattice exhibits, in frozen and displaced form, a bundle of social relations. Where this much has been recognized, it has frequently resulted in the 'problem' of how to map from one into the other. This is just a variant of the form–process puzzle, and as such continues to attract attention only because it has been incorrectly posed in the first place. It presumes that it is possible to identify spatial structure independently of social structure – this is a condition of its existence – whereas the real problem, if we accept the arguments of Hindess and Hirst, turns on the need to recognize (a) that spatial structures cannot be *theorized* without social structures, *and vice versa*, and (b) that social structures cannot be *practised* without spatial structures, *and vice versa*. The second of these is unexceptional, but it depends on and is inextricably bound to the first proposition.

It also follows, I think, that these remarks remove the need to turn to Piaget's operational structuralism, as Brookfield (1975) or Sayer (1976) would have us do. Indeed, I shall want to put this more strongly in a moment. Brookfield's reasons for attempting to do so are twofold: he is not prepared to exempt 'structures of the economic base' from 'the laws of continuous, interactive transformation', and he is unwilling to accept 'the strong elements of an *a priori* functionalism' which he finds in the classical Marxian model (Brookfield 1975, 196). Both of these conclusions are the result, as he admits himself, of a study of Harvey's writing in *Social Justice and the City* rather than an exegesis of Marx's own contributions, much less a critique of contemporary Marxism. We have already seen that Harvey's constructions are defective and that a more rigorous concept of social formation is capable of answering both of Brookfield's objections.

But we must go further. Sayer's (1976) critique of regional science makes a distinction between the structuralism of Piaget and Althusser, and locates its reformulation as an urban and regional political economy in the first of these problematics. The position is in fact a good deal more complicated than Sayer admits.

In the first place, there is a major correspondence between the two, inasmuch as Piaget (1971, 126–7) explicitly acknowledges Althusser's concept of structural causality. Hence it is no great surprise to find Sayer (1976, 248–9) emphasizing that 'spatial organization is an important problem for all economies regardless of whether they are feudal, capitalist or socialist, and yet the particular forms of spatial organization will *reflect* the different types of socio-economic relations encountered' and that 'the search for "*determining relations*" does not necessarily entail an infinite regress: the regress ends at the level of macro-political economy' (italics added).

In the second place, and connected to this, Piaget develops his 'authentic' structuralism in such a way that his critical intentions, which are shared by Sayer, and in particular his belief that 'man can transform himself by transforming the world and can structure himself by constructing structures' (Piaget 1971, 119), are severely compromised. He says that there is no need

to choose between the primacy of the social or that of the intellect; the collective intellect is the social equilibrium resulting from the interplay of the operations that enter into all cooperation. Nor does intelligence precede mental life or the reverse; it is the equilibriated form of all cognitive functions. And the connections between intellect and organic life may be conceived of in the same way: though it would not do to say that all vital processes are 'intelligent', it can be maintained that in the logical transformations which d'Arcy Thompson studied more than a generation ago life is geometrizing; today we may go so far as to say that in many respects life works like a . . . 'general' intelligence [Piaget 1971, 114].

'Ultimately', therefore, structures are no more and no less than 'logico-mathematical models' (Piaget 1971, 98). The affinities with modern geography are very clear: the cybernetic impulses of systems analysis are invoked; the geometric tradition is upheld; and mathematical model-building is vindicated.

It is hard to see, on this basis at least, how operational structuralism is much of a threat to the intellectual *status quo*, and Sayer (1976, 193) confesses that much of his critique may seem 'pitched too much on conventional regional science's own terms'. The cost of doing so is, I suggest, prohibitive. What Piaget's epistemology does is to resolve the relations between what we have called spatial structure and spatial schema in favour of the latter and, as Hindess (1977, 161) notes, 'the effect of this reference back to the structure of the thinking apparatus (the cerebral cortex) can only be to obscure the significance of the mathematical materials and theoretical constructions necessary to the formulation of a given mathematical result and of the conditions which determine that this result is or is not valid'. In this sense Sayer's conclusions are inconsistent with the incisive critique which precedes them, since his examination of the discourse of regional science is conducted in precisely these terms. In short, then, operational structuralism slides inescapably into a sophisticated dogmatism, and its problematic fails to specify – indeed, systematically occludes – the connections between theoretical construction and practical transformation. These must have a central place in any structural explanation which is to form part of a genuinely critical science.

4 Reflexive explanation in geography

The image, it is clear, must be set between the mind or senses of the artist himself and the mind or senses of others. If you bear this in memory you will see that art necessarily divides itself into three forms progressing from one to the next. These forms are: the lyrical form, the form wherein the artist presents his image in immediate relation to himself; the epical form, the form wherein he presents his image in mediate relation to himself and to others; the dramatic form, the form wherein he presents his image in relation to others.

James Joyce: *Portrait of the Artist as a Young Man*

Modern geography's first encounters with reflexive explanation were deeply felt but superficially conducted. They were prompted and sustained by geography's traditional attachment to particular places and the people that live in them, by what Harvey describes as the 'thread to geographic thinking which, at its best, produces an acute sensitivity to place and community, to the symbiotic relations between individuals, communities and environments' (Harvey 1974, 22). But valuable and vital as this was, it was liable to be snapped at any time, because it was never bound into a strong epistemological backcloth. In drawing away from a rigorous examination of the discourses which were involved in its project, therefore, geography inevitably resigned itself to what Harvey also calls a sort of 'parochial humanism', a set of descriptions which stopped short of a fully critical capacity to 'understand our condition and to reveal the potentiality for the future imminent in the present' (Harvey 1974, 24). I shall want to say that it is of the first importance to couple this long-standing concern with the structures of the social world to an equally fundamental concern with the conditions for speaking about them: in other words, to connect ontology to epistemology.

According to Relph (1976, 126), the contemporary landscape has seen an erosion of the specificity of the regions which have always engaged the attention of geographers: places have been deformed and made into arenas which respond to the external dictates of an abstract rationality rather than to the intentional structures of the people who live in them and invest them with their own meanings. And according to Gouldner (1976, 48), ideologies can be regarded in much the same way, since they speak of the world 'in an omniscient voice, as if the world itself rather than men were speaking': discourse becomes deformed and made into a contest rather than an exchange, in which the ideologue seeks to enforce a 'compulsively one-sided image of the world'. The link between the two is a close one, and Relph warns that geography is abandoning itself to a technically constituted ideology which contributes to the extension and reification of this, so to speak, compulsively one-surfaced, placeless image of the world. In doing so, he argues, geography is running the risk of 'losing contact with its sources of meaning', with the attitudes, intentions and experiences which confer particular identities on the realms it studies, and so ultimately of destroying the very richness which called it into being in the first place.

The dilemma is not a new one. Dardel (1952, 133) maintained that geography always has been and always will be 'stretched between knowledge and existence', between an environmentalism where the geographer 'seeks meaning in order and finds a largely determined, timeless and tidy world' and an existentialism where he 'seeks meaning in the landscape as he would in literature, because it is a repository of human striving' (Tuan 1971, 184). The difference between this second proposition and the claim discounted in the previous chapter, that the landscape somehow has the key to our understanding of it within itself, is, I hope, obvious. It amounts to the counter-claim that the key is to be found somewhere within ourselves, so that, by extension, an authentic – Tuan would say a humane – geography would not seek to impose its own (negotiated) frame of reference on the world but would instead attempt to understand other frames of reference and to *mediate* between what we might think of as various 'lay' and 'technical' constructions of the world. This chapter is concerned with this process of mediation, and serves to reinforce the earlier argument for an examination of discourse. But it does so in different terms, because the intention is now not simply to clarify the constructs of social scientists but also to interpret the constructs routinely drawn upon by the social actors themselves.

While this means that I am disposed to accept Schutz's (1962, I, 62) view that 'all scientific explanations of the social world can and, for certain purposes, *must* refer to the subjective meaning of the actions of human beings from which social reality originates' it does not follow that I regard this kind of appraisal as being in any way sufficient. Implicit in the problematization of discourse is the belief that there are constraints on social action which are so much part of the taken-for-granted life-world of the actors that they are either unable to verbalize them or would consider it irrelevant to do so. It must be part of the task of a critical science to disclose the constitution of these structures, and this has to involve going beyond the moments of a purely reciprocal contemplation which might, as Buttimer (1976, 89) suggests, help to 'elucidate how moorings in past experiences can influence and shape the present' but which inevitably has 'little to say about future directions'. The philosophical tradition which is involved in this is many-stranded, however, and a number of its threads can be woven into or even unravelled from other traditions; although geography's reflexive critique started out by concentrating on the propositions of phenomenology, therefore, it is important not to lose sight of these background convergences. They are only placed at the back of the stage by an artifice of lighting, as it were, and this will subsequently have to be changed if the limitations of a purely reflexive geography are to be recognized and ultimately transcended.

The need to do this is made much clearer once we attempt to connect the philosophical argument to the conduct of geographical inquiry. It is relatively easy to find concepts within the recent history of the subject which on the surface at least, seem to allow for some kind of reflexive explanation, and it will be part of my task to establish their grounding. Even so, it is only much more recently that discussions of this sort have been informed by an explicit consideration of debates within philosophy and, as Billinge (1977, 66) notes, we have to resist the *a posteriori* justification of these early ventures to be found in many contemporary commentaries. But perhaps more important, as Billinge also notes, the extent to which any of these recent discussions really have been – *or ever can be* – 'informed' by Husserl's phenomenology in particular is extremely problematic. There is an essential difference between the contemplative intentions of his transcendental philosophy and the practical concerns of a social science, so that it is scarcely surprising that where geographers have aligned themselves with Husserl's project their efforts have been directed towards the destruction of

positivism as a *philosophy* rather than the construction of a pheno-
menologically sound *geography*.

This is clearly not enough, and in this respect it is important to
consider the work of Alfred Schutz who, it has been claimed, at
times endorsed Husserl's transcendental philosophy and appeared to
operate within its confines, but at others was severely critical of
the bracketing of the 'natural attitude', our taken-for-granted pre-
suppositions about the world, which took place in its *epoché*.
Schutz objected to Husserl's almost obsessive and at times exclusive
concern with uncovering the 'essences' of social phenomena, and
proposed instead to investigate the way in which they were concealed by skeins of inter-subjectively woven social meanings. Where
Husserl had treated the ordinary assumptions we make in our everyday life as 'just so much bric-à-brac that has to be cleared away in
order to reveal subjectivity in its pure form' (Giddens 1976, 25),
therefore, Schutz (1962, I, 62) insisted that 'the most serious question
facing the social scientist is how to form objective concepts and
objectively verifiable theories of [these] meaning structures'.

In saying this Schutz was not simply substituting a constitutive
phenomenology of the natural attitude for Husserl's transcendental
phenomenology, inverting Husserl's *epoché*, as Giddens (1976, 27)
describes it, he was also accepting some (and only some) of the
propositions of Hempel. Thus, the statements made by any science,
natural or social, had to involve 'the explicit formulation of determinate relations between a set of variables', relations which had to
be 'capable of being verified by anyone who is prepared to make
the effort to do so through observation', so that 'a fairly extensive
class of empirically ascertainable regularities can be explained'
(Schutz, in Bernstein 1976, 137).

It is through background convergences of this sort that the
foundations for a set of reflexive explanations in geography might
be established, and Buttimer (1976) has in fact outlined a programme for a geography of the life-world which conjoins 'humanistic
and scientific enquiry' in a way which is broadly consistent with
Schutz's project. Once the concern for reflexive explanation is translated into this operationally more tractable context, the limitations
of a phenomenological geography ought to emerge more clearly,
and the stage will then be set for the illumination of hermeneutics.

This chapter is therefore divided into three main sections. The
first is confined to Husserl's critique of the modern European idea
of science and to what he sees as the necessity to dismantle it and

recover through the phenomenological reduction the classical conception of *theoria*. The second section narrows the focus and uses Schutz's work to link Husserl's treatment of the crisis of the European sciences in general to a geography of the life-world in particular, a project which its several authors intend – to paraphrase Habermas – to overcome geography's estrangement from what they take to be its legitimate task. Several difficulties will become apparent during this discussion, and the third section attempts to remove some of them by introducing the possibility of a deliberately hermeneutic geography, which, even in its present necessarily programmatic form, offers a more clearly defined critical orientation.

Husserl and phenomenology

Like any other social philosopher Edmund Husserl clarified, extended and revised his ideas throughout his life, but instead of trying to trace through the way in which he developed his transcendental phenomenology over the years I want to take up the story in 1935, just three years before his death. Hitler had already forced him to flee his native Germany, and by then he was in Prague to deliver what later became known as the *Krisis* lectures. These were clearly a response to the collapse of the Weimar Republic and the rise of Nazism, but they were also unequivocally directed against the naturalist conception of science which was being renewed through the logical positivism of the Vienna Circle. Husserl believed that the coincidence between the two was not an accidental one.

'The European nations are sick', he declared, and Europe itself was 'in a critical condition' which could only end either 'in the ruin of Europe alienated from its rational sense of life, fallen into a barbarian hatred of spirit; or in the birth of Europe from the spirit of philosophy, through a heroism of reason that will definitively overcome naturalism' (Husserl 1965, 150 and 192). It was barely a hundred years since Comte had denounced another great conflict which threatened to engulf the continent, and he too had taken refuge in the sanctity of reason. But whereas Comte had been close enough to the Enlightenment to believe in 'Reason with the capital "R" ' (Badcock 1976, 23), Husserl went right back to the classical conception of *theoria*. To the Greeks *theoria* had what Bernstein (1976, 175) calls an 'ultimate practical efficacy': it was charged with providing 'a "way" that intrinsically cultivates and educates the soul', releasing it from the confines of dogmatism and breathing the spirit of pure reason into the corpus of society. *Theoria* was a 'universal science', therefore, and brought about 'a progressive transformation that ultimately draws into its orbit the entire spiritual culture of mankind' (Husserl 1965, 163). Husserl argued that in the modern world this ideal had been severely compromised and its practical efficacy reduced to a narrowly technical domain. Science – and particularly positive science – had turned man into an object in an objectified world, and it was no longer aware of its origin in 'certain fundamental motives of the life-world itself' (Kockelmans 1967, 252).

This needs elaboration because it is absolutely central to Husserl's critique and his consequent faith in the ability of a transcendental phenomenology to resolve the crisis and restore 'the spirit to itself' (Husserl 1965, 189). His fundamental complaint was that the modern scientist 'does not make it clear to himself that the constant foundation of his admittedly subjective thinking activity is the environing world of life. This latter is constantly presupposed as the basic working area, in which alone his questions and his methodology make sense.' As we saw earlier, Habermas makes much the same point. Husserl wanted to know 'where, at the present time, is that powerful bit of method that leads from the intuitive environing world to the idealizing of mathematics and its interpretations as objective being, subjected to criticism and clarification?' (Husserl 1965, 185–6). Not in positivism or in any other form of naturalism, that much was certain: although Comte had explicitly grounded science in *le réel*, he had understood this in a much more restricted sense than that required by Husserl and had never examined the way in which man's experience of the world was reciprocally tied to his propositions about it. The need to do so had never even been acknowledged, since the possibility of cognition had been taken for granted and the procedures of positive science had been treated as replications of the natural attitude and hence as intrinsically nonproblematic. Husserl's view, by contrast, was that the major task of a rigorous philosophy ought to be to interrogate the connections between the constitution of meaning which takes place within the life-world and the process of objectification which takes place within science, so as to reveal the limitations of the naturalist enterprise, to establish the validity of other forms of cognition, and finally to reclaim for man an awareness of the total life-world which the strictures of contemporary science had denied to him.

Husserl argued that any exploration of the possibility of cognition must start by recognizing that cognitive experiences are 'object-constituting events': that they possess an *intentional structure* through which objects are made to mean something to us. All cognitive experiences are thus supposed to have a directional character, 'implying the active "movement" of consciousness beyond itself to include [i.e. "intend"] an object within its sphere', so that the object becomes merely a correlate of the act of cognition itself (Gorman 1977, 22). To simplify a complicated distinction, we can say that Comte's phenomenalism was an attempt to disengage from the intentional structures and fasten on the objects, while Husserl's phenomenology was an attempt to disengage from the objects and withdraw into the intentional structures. Through an act of pure philosophical reflection, the *epoché*, Husserl sought to suspend (bracket) the natural attitude in order to disclose what he described as 'a self-contained realm of being into which nothing can penetrate and from which nothing can escape' (Husserl, in Pivčević 1970, 77). There, free of all empirical particulars, the very essences, the *eidé*, of phenomena would stand revealed.

The essence of a phenomenon determined the confines within which it had to remain if its constitution were to stay the same, if it were to be intended as one object rather than another. Since essences were therefore by definition invariant, necessary and general, it followed that phenomenalism's denial of the *epoché* condemned it to the contemplation of the contingent, distracted by inessential empirical variations to such an extent that it could never achieve proper scientific status, while phenomenology's espousal of the *epoché* allowed it to claim universal validity as 'eidetic science'. (As I have argued elsewhere, the parallel with linguistic structuralism is a close one (Gregory 1978), and Husserl was a major influence on Jakobson and the Prague school of structural linguistics.) It was from the eidetic base that the ascent back into the everyday world was to be mounted. Once subjective consciousness had been experienced in this its 'pure' form, it became possible to re-emerge and reconstitute the real historical world 'in all its uncouth complexity' (Giddens 1976, 25).

The path to truth, which is how Husserl conceived of his project, thus led through successive disengagements from and re-engagements with the world, and this demanded an intense effort, 'comparable in the beginning to a religious conversion' (Husserl, in Bernstein 1976, 130). But he was convinced that nothing less could reclaim *theoria*.

E

It can be seen from this, I think, that as far as human geography is concerned the importance of Husserl's transcendental phenomenology is *not* that it provides for a subtle return to uniqueness, as has so often been claimed in the past. It was certainly *not* Husserl's intention to uphold the validity of a multiplicity of different frames of reference: on the contrary, he made it perfectly clear that the whole purpose of the *epoché* was to ensure that the world could be 'identically reconstituted in each separate individual through a similar reflective procedure' (Gorman 1977, 125), and in this sense his programme was motivated by an impulse every bit as critical and as objective as Comte's insistence on *le précis*. Where it differed from it, of course, was in its rejection of the determining force of universal laws, and it is only in these terms that it can be regarded as an invitation to idiographic inquiry.

But at the same time, it is not difficult to see how his phenomenology *has* been taken to provide the philosophical props necessary to support an excavation of purely individual 'geographies of the mind'. In the first place, Husserl was never very explicit about precisely how the transition from the empirical to the transcendental level was to be effected, and nor did he have much to say about precisely how the world was to be reconstituted once it had been put in brackets (Pivčević 1970, 73). In the second place, geography's interest in intentionality has been largely a descriptive rather than a prescriptive one, so that to accept Husserl's critique of positivism – and with it the importance of referring explanations to the subjective domain of the life-world – is not necessarily to attempt the transcendental exposure of 'subjective objectivity' which takes place in the *epoché*.

Where Husserl's work has been of particular importance, therefore, is in its vigorous rejection of the aspirations and assumptions of positivism. If it (necessarily) fails to provide an alternative which is congruent with more practical interests it has at least afforded a convenient rallying-point for many dissidents who, although they would undoubtedly regard a joint commitment to reflexive explanation as a very low common denominator indeed, have nevertheless been able to accept enough of the phenomenological critique to make a concerted thrust with it at the heart of positive geography. The wound has been a strategic one: 'it questions the meaning of our activists as scientists: does gaining an intellectual grasp of problems make us more sensitive to our world, or does it remove us from it? Can our research models and routine interaction within the profession promote a keener sensitivity to our own lived worlds or to those of other people?' (Buttimer 1976, 291). Questions like these have been raised many times in the past, of course, but almost invariably from within or at most as a peripheral extension to the positivist problematic, whereas Husserl poses them in such a way that they become not only inimical to any form of naturalism but also central to any authentic social science. In so far as positivism had brought 'hard-headedness' to the subject, then it evidently had to be countered by much more than an inchoate concern at geography as a science of spatial relations (Harris 1971), and Husserl's phenomenology came to represent a peculiarly forceful statement of the limitations of a geography divorced from the life-world.

In some ways the debate has been a re-enactment of the original encounter between Schlick and Husserl in the 1930s, but an important difference – however paradoxical – is that the current allegiance to what Husserl might have described as the least pragmatic of all philosophies has been at root a pragmatic phenomenon. Its protagonists clearly accept elements of the phenomenological critique, and have translated some of their intentions into its terms so successfully that the hold of positivism on geography has been markedly weakened; but they are unlikely to find an acceptable realization of these intentions within Husserl's problematic. Hirst (1977, 79) has shown that 'Husserl's position is structured around a central and inescapable contradiction between his theoretical position on the one hand and the terms in which that objective is to be realised on the other'. There is, I think, no need to discuss this here, since we can simply treat it as reinforcing the misgivings which were mentioned earlier: what is more important in the present context is that, as Entrikin (1977, 629) concludes, the phenomenological approach in geography 'is a form of criticism rather than an alternative to the scientific approach'.

The question that needs to be answered, then, is whether this strategic offensive – which, to repeat, is not without its own epistemological difficulties – can be turned into a practical victory without distorting the basic forms of reflexive explanation.

Towards a geography of the life-world

In *Explanation in Geography* Harvey rejected the possibility of a scientific geography of the life-world. This would have to involve according primacy to subjective experience, and would inevitably restrict explanations

> to a statement of the intentions, reasons, motives and dispositions involved in a given act and such a restriction also limits the kind of question that the social scientist and the historian can ask. At this level of analysis it is difficult (especially in historical situations) to envisage any other form of validation than *verstehen* [i.e. 'imaginative understanding'], and given that *verstehen* forms an exclusive and exhaustive form of validation, it is difficult to envisage explanations ranging beyond a statement of intentions, dispositions, motives and reasons. This vicious circle is difficult to break out of [Harvey 1969, 56].

The result was what he subsequently called a formless relativism, and he concluded – and he was by no means alone in this – that although *verstehen* 'may be fundamental to hypothesis formation' it could not in itself constitute the foundations for a methodologically secure geography.

Harvey was not trying to burn a straw man. The exhortations of Sauer and Wooldridge to see the land with the eyes of its own occupants had been brought together and generalized in Wright's (1947, in 1966, 83) *geosophy*: 'to geography what historiography is to history, it deals with the nature and expression of geographical knowledge, both past and present – with what Whittlesey has called "Man's sense of terrestrial space" '. It embraced 'the geographical ideas, both true and false, of all manner of people' – not just geographers – and so it had to deal 'in large degrees with subjective conceptions'. If Wright had had little to say about how these were to be articulated the import of his message was clear enough, and it was subsequently amplified by Lowenthal's (1961, 260) lucid reminder that 'we are *all* artists and landscape architects, creating order and organising space, time and causality in accordance with our apperceptions and predilections' (my italics). It was obvious to Lowenthal that 'agreement on such subjects is never perfect nor permanent', so that geographers ought to expect 'only partial and evanescent concordances' in their explanations. In short, what was seen from within the positivist problematic as a severe problem of scientific validation was seen from within the *verstehende* tradition as a clear mandate for geography as an essentially artistic enterprise: 'environmental experience is most fully apprehended, we suggest, when the otherwise banal or amorphous circumstances of everyday existence are enlarged and transformed through the medium of mystery and art' (Lowenthal and Prince 1976, 130). The main target of Harvey's criticism, behind these various geographical interpretations, was Max Weber. He can scarcely be called the father of *verstehende* social science, however, and the title more properly belongs to Wilhelm Dilthey. I make this point because whereas Dilthey had regarded *verstehen* (understanding) and *erklären* (explanation) as two quite separate procedures, Weber always connected them closely together. 'To be able to put one's self in the place of the actor is important for clearness of understanding', he said, 'but not an absolute precondition for meaningful interpretation', because 'every interpretation strives to achieve utmost verifiability' (Weber, in Freund 1968, 99). Since this could not be achieved entirely within the realm of subjective meaning, it was necessary to construct an 'ideal type', an extreme rationalization, to project intentions on to the actions through which attempts were made to realize them and so gain entry into a realm in which empirical validation could take place. In this way the ideal type was supposed to provide a means of moving between *verstehen* and *erklären.* Guelke (1974, 199) has suggested a similar, although highly programmatic, procedure for an idealist geography, which seeks 'to understand the development of the earth's cultural landscapes by uncovering the thought that lies behind them' through the construction of 'ideal models' which 'map out rational courses of action' which can then be empirically validated. What all this indicates, of course, is that although Harvey was right to underline the difficulties encountered by an interpretative social science, he was mistaken to suppose that Weber was unaware of them.

Now, Alfred Schutz argued that the success of Weber's solution depends on what we understand by 'subjective meaning', because until we know to what this refers we have no criteria for recognizing its objective expression in the world of social action and hence no real means of empirical validation at all. The missing link, he contended, had to be forged in a phenomenology which took the prolem of meaning as its starting-point, and it was for this reason that he turned to Husserl (see Gorman 1977, 16–20). It is essential to recognize that the place and function of Husserl's phenomenology in Schutz's scheme were 'prescribed and precisely determined by the nature of the basic project which Schutz found in Weber, namely to reduce "the world of objective mind to the behaviour of individuals" ' and that Schutz made 'no attempt to establish that the object of his proposed science is a possible or coherent object of

investigation *or that its definition is compatible with the funda-*
mental concepts of Husserlian phenomenology' (Hirst 1977, 57–8;
my italics). Schutz was easily the most important broker between
the Weberian and the Husserlian schools – among the others, incidentally, was Alfred Weber, whose *Kultursoziologie* embraced
elements of both his brother's sociology and Scheler's phenomenology (Hawthorn 1976, 183) – but the price he had to pay for the
retention of Weber's commitment to empirical validation was the
rejection of Husserl's transcendental *epoché*. For all the willingness
with which Schutz made the sacrifice, it had awkward consequences
for his projected social science. In what follows I want to spell
these out through a discussion of the twin pillars of his so-called
phenomenology of the social world, meaning and validation, and
at the same time to explore their implications for a geography of
the life-world.

Meaning, intentionality and the life-world

Schutz's account of subjective meaning can be summarized in terms
of three linked propositions.

First, he says that there are two ways in which 'I' can know the
world.

On the one hand I can look upon the world presenting itself to me as one
that is completed, constituted and to be taken for granted. When I do this,
I leave out of my awareness the intentional operations of my consciousness within which their meanings have already been constituted.

The Husserlian legacy is obvious: this is the pre-reflective world
of the natural attitude.

On the oher hand [he continues], I can turn my glance toward the
intentional operations of my consciousness which originally conferred
the meanings. Then I no longer have before me a complete and constituted world but one which only now is being constituted and which
is ever being constituted anew in the stream of my enduring ego: not
a world of being, but a world that is at every moment one of becoming
and passing away – or better, an emerging world. As such it is meaningful to me in virtue of the meaning-endowing acts of which I become
aware by a reflexive glance [Schutz 1967, 35–6].

Schutz is claiming here that *only the already experienced is mean-*
ingful, because 'meaning is merely an operation of intentionality'

which 'only becomes visible to the reflective glance' (Schutz 1967, 52).

Secondly, this allows him to claim that although social actions are oriented to the future they are executed within a framework which is determined by these reflective, retrospective glances: they are 'projects' in both senses of the term. As such, they are organized through a *system of typifications*, sets of constructs founded in our unquestioned past experiences which allow us to anticipate the meanings of our future encounters with the world and which may themselves be changed as a result of these engagements.

Thirdly, Schutz wants to claim the existence of *socially approved* systems of typifications, so that although 'we define the common-sense world as unique subjects' we nevertheless 'think in typically familiar patterns and act in typically familiar ways' (Gorman 1977, 44). Our frames of reference are thus negotiated with respect to our past experiences and with respect to one another.

Each of these propositions has its counterpart in geography, but most of them are so ill-defined that I will mention only two of them here. Lukermann (1964) was among those who recognized the importance of the historicity of meaning; he argued that man's past environmental experiences and beliefs underlie his day-to-day actions, so that the 'character of place' must always be 'emergent and becoming' and geography must be approached historically. The typification of meaning and the part society as a whole plays in it were most obviously elided in Kirk's (1951; 1965) Behaviourial Environment, where 'the same empirical data may change itself into different patterns and have different meanings to people of different cultures, or at different stages in the history of a particular culture'. These culturally defined structures of meaning, which Kirk called *gestalten*, were incorporated into sets of typical (community) responses to changes in the external or Phenomenal Environment (Figure 11). Geography was therefore obliged to enclose this 'psycho-physical field' since it was within its confines that 'rational human action begins'.

It would be wrong to pretend that either of these parallels, or any of the others which might be drawn, were in any way direct translations of Schutz's propositions. As Billinge (1977, 59) observes, in all cases their understanding of intentionality was an extremely loose one; Lukermann's argument resembled Vidal's incision of custom into landscape as much as it did Schutz's world of becoming and passing away, and Kirk's Behavioural Environment was fashioned from Wertheimer's psychology rather than Schutz's phenomenology. But despite their diverse ancestry it is, I think, legitimate to say that there were important *immanent* relationships between them, and that it was the failure to acknowledge these which prevented the development of a conception of meaning which came anywhere near the completeness of Schutz's construction. This is something of a two-edged sword, however, because in many ways it is the looseness of their formulation which allows them to be retrospectively cast in Schutzian terms, conjoined from the

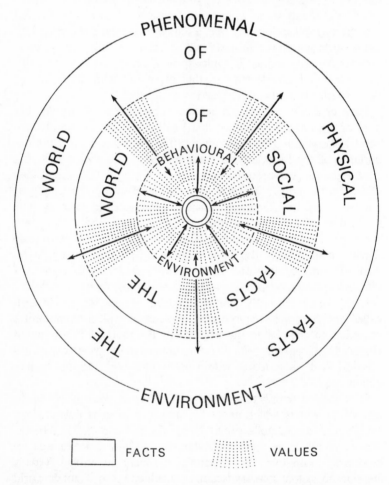

FACTS   VALUES

Figure 11 *Behavioural and Phenomenal Environments (after Kirk)*

outside as it were, while at the same time it was this lack of precision that blurred their edges and prevented them from connecting up on the inside. Their separate advocacy did not result in a concerted exploration of either their assumptions or their implications, and it was not until comparatively recently that geographers cautiously, and by no means unanimously, came to recognize the possible significance of an explicitly phenomenological appraisal of meaning.

As in the case of Husserl, a major preoccuption of Schutz's had been the derivative status of the separate sciences, and in much the same way that he wanted to expose their roots in the life-world and to investigate the intersubjective constitution of meaning which they simply took for granted, so geography began to uncover its foundations in what Lukermann had previously identified as the character of place. If modern geography had been built on this, then it was essential to know 'what *is* this sense of place on which we have not only erected a spatial geography of considerable elegance but, more important, on which we still depend for the decisions and acts in our daily lives?' (Tuan 1975, 246). The question was a highly charged one. As I have already said, an interest in the 'spirit', the 'meaning', the 'sense' or even the 'personality' of a region had always been a traditional concern of the subject – the way in which a place 'through long associations with human beings can take on the familiar contours of an old but still nurturing nanny' is how Tuan put it – but he was not asking for an evocation of these unique/collective attachments: he wanted to know how these constellations of meaning were *possible*, and on two related levels. The first-order task was to discover the way in which individual actors constitute the life-world in general and places in particular, and the second-order task was to discover the way in which geographers typically constitute these lay constructions and incorporate them into their own accounts.

These inquiries met with resistance not only from positivist geographers who (understandably, if erroneously) complained that they had nothing to do with the proper conduct of science, but also from those who had never been persuaded by the tenets of positivism but still could not see that they had anything to do with the proper conduct of geography. The publication of Tuan's *Topophilia* (1974) and Relph's *Place and Placelessness* (1976) did little to remove their uncertainty: the style and content of both of them were so unfamiliar that, as Harris (1975, 163) observed, they 'opened up a

world that has been as remote from most of our thoughts as, physically, our predecessors were remote from the source of the Zambezi'.
But, as Relph (1976, 6) pointed out, 'while place is often considered
as a formal geographical concept, any exploration of place as a
phenomenon of direct experience cannot be undertaken in the terms
of formal geography nor can it solely constitute part of such geography. It must, instead, be concerned with the entire range of
experiences through which we all know and make places, and hence
can be confined by the boundaries of no formally defined discipline.'
In other words, these investigations demanded that geography
recognize the *totality* of subjective experience which it normally
took for granted and out of which it objectified its systems and
structures. In doing so, it would have to accept the fundamental
importance of intentionality and see action and interaction as 'a
response to the meanings which are attached to and define the
social and spatial institutions against which the action is set' (Cullen
1976, 406). A geography of the life-world would then be possible,
but any privileged status for it would be inadmissible.

The elements of such a geography can now be presented in summary form, broadly consonant with Schutz's phenomenology, but
it has to be remembered that these materials are only provisional:
we are still a long way from providing a satisfactory answer to
Tuan's question, and each of these propositions contains its own
difficulties.

(1)   We become aware of meaning-endowing acts through spatial
as well as temporal distance: 'it is possible to be aware of our
attachment to place only when we have left it and can *see* it as a
whole from a distance' (Tuan 1975, 225).

(2)   Hence the geography of our life-world emerges out of our
various interactions and engagements as 'a kind of topological surface punctuated by specific anchoring points', each one 'stamped
by human intention, value and memory' (Buttimer 1976, 283–4).

(3)   But we do not have to encompass a multiplicity of entirely
individual geographies, since for the most part our experience of
the world is routinely accepted and codified through a set of social
typifications: according to Buttimer (1976, 286), 'the key message
of phenomenology for the student of social space is [thus] that much
of our social experience is pre-reflective: it is accepted as given,
reinforced through language and routine, and rarely if ever has to
be examined or changed'.

(4)   And since, as Cullen (1976, 406) insists, the existence of these typifications depends on their being continually reaffirmed in social actions these must conform to and be revealed in typical, predictable space–time rhythms.

(5)   *A geography of the life-world must therefore determine the connections between social typifications of meaning and space–time rhythms of action and uncover the structures of intentionality which lie beneath them.*

As Duncan and Sayer (1977) have argued, however, a major deficiency of the programme as it stands is its restricted conception of social structure: in particular, it ignores the material imperatives and consequences of social actions and the external constraints which are imposed on and flow from them. These limitations cannot be overcome within Schutz's problematic, and Buttimer (1976, 290) accepts the need 'to move beyond the letter of the phenomenological law' – to break out from what King (1976, 305) describes as a world 'in which self-reflection is esteemed above all else' – and to uncover 'the dynamics of processes already operative which set the rhythms of time and space for everyday life situations'. And this might be achieved most satisfactorily, she argues, within the framework provided by Hägerstrand's time-geography.

Hägerstrand's starting-point is that 'human beings are always seeking to reach goals, some immediate and some more distant, some of an individualistic and some of a collective nature'. This means that it must be possible – *for the observer as well as for the actor* – to 'group events into coherent clusters, each cluster [or project] representing the steps necessary for movement towards each goal'. If space and time are viewed as resources which the individual has to draw upon in order to make these moves, then the various projects can be formed into nested hierarchies in space–time. Then, 'each project would seem to try, between its beginning and its end, to accommodate its parts, be they tangible or not, in the surrounding maze of free paths and open space–time left over by other projects or gained through competition with them' (Hägerstrand 1973, 78–9).

The introduction of competitive allocation is important because it prepares the way for an examination of the asymmetries of power which are markedly absent from Schutz's phenomenology (Giddens 1976, 53). So far, Hägerstrand has identified three basic kinds of constraint – capability constraints, coupling constraints and authority constraints – but these have all been derived empirically and they clearly ought to be tied to concepts derived from a properly constituted social theory, which would then allow for a more rigorous incorporation of power into the problematic. And Häger-strand himself admits that there is an urgent need for a way of 'dealing with power in space–time terms of considerable precision': power relations are of such immense importance for the under-standing of how projects compete in the available space–time that their analysis could develop into the core of a new human geography (Hägerstrand 1973, 85).

Its object would be to uncover 'structural patterns and outcomes of processes which can seldom be derived from the laws of science as they are formulated today' (Hägerstrand, in Pred 1977, 211). The failure of modern science, he argues, is a result of its neglect of synchorization and synchronization: the inescapable necessity for space–time 'packing' in the conduct of practical life. Hägerstrand ties the structures of the social and the natural worlds into the organization of space–time through the identification of *paths*, which define, via coupling constraints, 'where, when and for how long, the individual has to join other individuals, tools and materials in order to produce, consume and transact', and of *domains*, which define, via authority constraints, systems of regulation which govern the actions of particular individuals or groups (Hägerstrand, in Pred 1973, 39–42) (Figure 12).

Now, although Hägerstrand's work is often described as a theory of society or a 'socio-ecology' (Pred 1973; Thrift 1977a), it should now be clear that it is not, at least not yet, and that it requires development in two principal directions, both of which are latent within its present formulation. First, there is a *convergence* between Hägerstrand's (still schematic) conception of paths and domains and the conception of a spatial implication within the social formation discussed in the previous chapter. These connections need to be strengthened, since they ought to clarify the relations between the various constraints and, in particular, to enable the theorization of the way in which movements through the space–time block are dialectically tied to the reproduction and transformation of social structures. Secondly, there is a *complementarity* between Häger-strand's conception of a project and Schutz's conception of a project since, as we have seen, Schutz is not much interested in their realization (which would necessarily intrude on asymmetries of power) and Hägerstrand is not much interested in their constitu-tion. Indeed, Hägerstrand regards projects as 'empirically given,

(i) ISOTROPIC SPACE-TIME

time

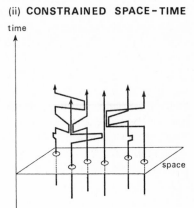

space

(ii) CONSTRAINED SPACE-TIME

time

space

(iii) PATHS AND DOMAINS IN SPACE-TIME

time

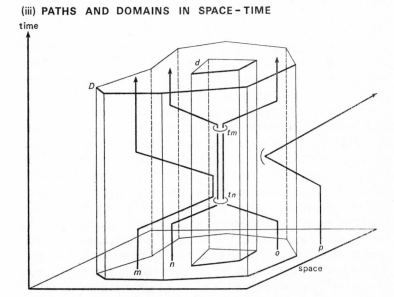

space

Figure 12 *Projects and constraints in space–time (after Hägerstrand)*

pre-existing structures' (van Paassen 1976, 339), so that their most important feature as far as he is concerned is their 'sequential order; sometimes manifesting itself in terms of very fixed intervals, sometimes flexible enough for survival despite time-lags' (Hägerstrand 1973, 79). Schutz also recognizes that, to continue in Hägerstrand's vocabulary, projects of different life-spans interlock within and between hierarchies, but he is much more involved in the way in which this can be related to constantly shifting systems of relevance and typification (Giddens 1976, 29). This tie – one is almost tempted to say umbilical cord – must be retained if we are to avoid what Giddens (1976, 156) terms a conceptual 'cutting into' the continuity of action.

I cannot develop these propositions in the present text, but, taken together, they insist on 'an interplay between subjective intentionality and the universe of external objective social relations', and in accepting this, reflexive explanation is confirmed as an essential moment in a critical geography whose 'imperatives lie in the revelation and pursuit of emancipatory and yet feasible lines of action at any given concrete historical conjuncture' (Scott 1976a, 635).

Validation, adequacy and the life-world

Schutz refused to take Husserl's transcendental turn because he believed that the propositions of social science ought to be capable of empirical validation. The primary contribution of phenomenology was to be its clarification of meaning, and once this had been achieved it would be possible for the second-order constructs of the social scientist to be made congruent with the first-order constructs (meanings) of the individual social actors. 'Each term in a scientific model of human action must be constructed on such a way that a human act performed within the life-world by an individual actor in the way indicated by the typical construct would be understandable for the actor himself as well as for his fellow-men in terms of commonsense interpretation of everyday life' (Schutz 1962, i, 44). Schutz referred to this notion of congruence as one of adequacy; but he never established how it was to be achieved, and Giddens complains that it is not at all clear what he meant by it. It is, furthermore, a nonsense even on his own terms: the *epoché* was founded on a problematization of social constructions of reality which were supposed to be *beyond* the reach of immediate consciousness, and without entering into a transcendental realm it is hard to see how the second-order constructs of the social scientist can ever achieve the apodictic certainty which would ensure a recognition by and a replication in the first-order constructs of the social actors themselves.

The reason for this, as I shall say, *necessarily* imprecise image is to be found in what Gorman (1977, 139) calls Schutz's 'dual vision' of an objective science of subjective knowledge. Husserl had brought the two together at the transcendental level, but this way out was not open to Schutz. As a result, 'by abandoning the transcendental ego and its correlative absolute knowledge, and simultaneously accepting Husserl's belief in the supremacy of individual, subjective consciousness in formulating the meaning each person experiences, Schutz has apparently eliminated the only element in Husserl's scheme that is in any way "objective" ' (Gorman 1977, 33; see also Entrikin 1977, 630).

This makes it extremely difficult to think of validation in anything other than individual terms and appears to consign social science to a realm of purely private interpretations and beliefs. Each one of us constructs the world in his own way, whether as actor or observer, and we have no means of determining the success of translations from one frame of reference to another. Just as Archimedes was unable to find a place from which to move the earth, so we are unable to establish a universal empirical basis from which to assert the truth of our individual claims to knowledge.

In practice, of course, this is not entirely so. As we have seen, Schutz maintained that our projects are organized through socially approved systems of typifications: 'we assume if we act in ways typically similar to previous actions, under typically similar circumstances, we will bring about typically similar states of affairs. This implies that we also ideally typify the course-of-action motives of those we interact with' (Gorman 1977, 54). *But if this is true of the life-world in general it is also in some sense true of social science in particular.* For all its deficiencies, Kuhn's (1970) account of the way in which paradigms are reciprocally formed around groups of scientists suggests a similar process of communal approval and typification. It is here that the coupling of what I earlier called 'lay' and 'technical' constructions of the world assumes strategic significance. The point is that validation is not simply a question of how to get from one enclosed, 'lay' frame of reference to another, equally enclosed 'technical' frame of reference, but rather that all such frames, lay or otherwise, are of necessity mediated by one another. 'The process of learning a paradigm or language-game as the expression of a form of life is also a process of learning what that paradigm is not: that is to say, learning to mediate it with other, rejected alternatives, by contrast to which the claims of the paradigm in question are clarified' (Giddens 1976, 144). This turns the problem into a hermeneutic one, and few geographers have bothered with it in these terms. In so far as they have acknowledged the interpretative tradition at all, whether in its Husserlian, Weberian or Schutzian forms, geographers have generally taken a 'consensus' of actors or observers as a sufficient vindication of the authenticity of their reconstructions, while at the same time admitting the possibility of 'alternative geographies'. In failing to specify precisely how such a consensus is negotiated, of course, geography runs with both the hare and the hounds. Thus, for example, Guelke (1974, 202) claims that the goal of an idealist geographer 'is to provide a true account and explanation', but admits that in practice this is unattainable since 'different interpretations can often survive quite happily because of the lack of data and the difficulties in precisely inferring an agent's intentions'.

But Bartels (1973, 30) argues that this sort of response is as inadequate as that of the opposing, instrumental tradition – which refuses to recognize the problem at all – inasmuch as neither of them incorporate a theory of cognition into their process of validation. As Baker (1977a) observes, at every step in accounts of this kind one is forced to ask 'How do you know?' and yet few geographers have seriously tried to white-line their limits of inference in a general, hermeneutic way. To remedy this definitely does *not* involve laying down a set of standardized interpretative procedures. The task of hermeneutics 'is not to develop a procedure of understanding but to clarify the conditions in which understanding can take place' (Gadamer 1975, 262), so that we might rephrase the original question and ask exactly *how* one frame of reference mediates another.

Hermeneutics and geography

As Ricoeur (1965, 26) wryly observes, the historical–hermeneutic sciences aim to disclose coherence within a frame of reference which at first seems to be incoherent in some way, and yet they are themselves in a state of considerable confusion. It is important not to minimize these internal differences, but here I simply want to establish the consequences of two of the less contentious of their principles. Both of them have been mentioned already: (a) any interpretation moves within a hermeneutic circle and (b) any interpretation changes that which is interpreted.

Taken together – and, as Pinkard (1976) shows, they have to be taken together – these propositions confirm the impossibility of any absolute standard of adequacy. Ultimately, says Taylor (1971, 14). 'a good explanation is one which makes sense of the behaviour; but then to appreciate a good explanation one has to agree on what makes good sense; what makes good sense is a function of one's readings; and these in turn are based on the kind of sense one understands'. When Guelke (1974, 202) claims that an idealist interpretation involves sorting out the intentions of actors 'in such a way that their actions can be seen as rational responses to their situations as they saw them', therefore, he glosses over *who* is to regard them as rational and on *what* basis. There can be no clearly defined distinction between understanding and explanation. On the contrary, as Bernstein (1976, 166–7) observes, what we take to be genuine first-order intentions, and consequently what we regard as adequate second-order representations of them, is strategically dependent on the substantive theory of motivation and rationality that we accept. In short, 'what we judge to be an adequate interpretation of social action is itself dependent on our understanding of the causal determinants of social action': that is, on the constructs, both lay and technical, which are embedded in our *own* frame of reference.

It must follow from this that interpretation is not a process of somehow *overcoming* the distance between one frame of reference and another. Gadamer (1975, 264) saw this as the cardinal error of historicism, the assumption 'that we must set ourselves within the spirit of the age, and think with its ideas and thoughts, not with our own, and thus advance towards historical objectivity'. The issue is obviously not confined to historical interpretation *per se*, although historical geography has admittedly produced more advocates of this assumption than any other branch of the subject. However, and as Gadamer went on to say, 'the important thing is to recognize the distance in time as a *positive* and *productive* possibility of understanding' (my italics). Interpretation certainly requires what Giddens (1976, 149) calls 'immersion', therefore, but this does not and cannot involve the substitution of one frame of reference for another: 'For what do we mean by "placing ourselves" in a situation? Certainly not just disregarding ourselves. This is necessary, of course, in that we must imagine the other situation. But into this situation we must also bring ourselves. Only this can fulfil the meaning of "placing ourselves" ' (Gadamer 1975, 271).

Most geographers would surely agree with Andrew Clark's (1962, 233) evocation of 'the tremendous satisfaction in feeling that one has gotten under the skin of a region and has at his command enough detailed knowledge of it to be able to disentangle the warp from the woof of its fabric and to identify the most important individual threads of its character'. But I think it fair to conclude that the interpretative movements in geography to date have in effect and with very few exceptions served only to conceal the tension which must exist between one frame of reference and another, the deep resonances and discordances which are struck by superficially similar clusters of meaning, and that the hermeneutic task must be to make these explicit and to clarify what makes such a vital 'immersion' possible.

In doing so, too, and in reclaiming its traditional attachment to particular places and the people that live in them, geography will have to dismantle the oppositions between subject and object, actor and observer, and emphasize the mediations between different frames of reference. Such an 'epical geography', as Joyce might have put it, can therefore never close its accounts once 'the agent's goal and theoretical understanding of his situation have been discovered', as Guelke (1974, 197) would want, because a social science that limits itself to meaning is not fully reflexive at all. It fails to elucidate, to bring into our consciousness, the constitution of irredeemably *practical* structures, and for this reason Scott's (1976, 633) criticism of Olsson's project, *Birds in Egg* (1975), that 'by failing to situate human thought and action within a wider social and historical totality, it never effectively comes to terms with precisely those problems of individual human existence that it seeks so strenuously to elucidate', is clearly capable of a much wider application.

5 Committed explanation in geography

Theory isn't abstract; it isn't words on a page; it isn't . . . aesthetically pleasing patterns of ideas and evidence. Theory is concrete. It's distilled practice. Above all, theory is felt, in the veins, in the muscles, in the sweat on your forehead. In that sense, it's moral . . . and binding. It's the essential connective imperative between past and future.

Trevor Griffiths: *The Party*

Geography has always been committed in one way or another, and in this chapter I want to provide a critique of the way in which it has conceived of the relationship between what, for the moment, I will simply call theory and practice. This means that my discussion will not – directly – address the specific *forms* in which, over the years, geography has provided a stream of descriptions, models and recommendations which Harvey (1973, 150) characterizes as either *status-quo* or counter-revolutionary. This is clearly an important task, but it is necessarily implicated in the more general one of showing how, in *structural* terms, these consequences necessarily flow from theoretical formulations which claim a detachment from, or at most a contingent relation to, political practice.

King (1976) has drawn on Rein (1976) to identify three ways in which positive geography has acknowledged a connection between theory and practice:

(1) *Mapping*, which involves the establishment of a set of cause and effect relationships within the domain of social science and the specification of a set of means and ends relationships within the domain of private or public policy-making. A broad correspondence between the two sets is then assumed, so that it becomes possible to translate ('map') from one into the other (Rein 1976, 43–4).

(2) *Code harmonization*, which involves a recognition that 'society works with various legal, religious and moral codes that are to some extent independent of one another. Code diversity then

becomes a problem because each code has its own internal requirements, and these invariably produce conflicts and create insuperable contradictions.' If these conflicts can be exposed within the domain of social science then they can be removed within the domain of policy-making by harmonizing the codes (Rein 1976, 45–6).

(3)   *Feedback*, which involves 'measuring the gap between ideals and practice (evalution)' and 'examining alternative means for more effectively and efficiently narrowing the gap (experimentation)'. These monitoring processes take place within the domain of social science and then feed back into the domain of policy-making, where 'the established profile of power or some new power coalition is prepared to move from one situation to another in response to the facts it receives' (Rein 1976, 49–50).

All three of these procedures make the relationship between the two domains a matter for individual or, at most, corporate decision. Admittedly, the decision might not always be an entirely voluntary one since, as Berry (1970, 22) warned, 'if we, as geographers, fail to perform in policy-relevant terms, we will cease to be called on to perform at all', but the link remains a matter of tactical decision rather than strategic evaluation.

It soon became clear that such a tactic could easily turn into a pragmatism, even an opportunism, which was increasingly difficult to reconcile with the traditional conception of academic freedom. If this were to become simply the freedom to sell oneself to the highest bidder then, Prince (1971, 150) declared, geography would be – and in large measure already was – 'a serious game played by professionals, by hired mercenaries – but, nevertheless, just a game'. As Leach (1974, 42) succinctly put it, all research reflects the paymaster, and if geography had to be 'relevant' then it was only proper to ask 'relevance to what and for whom?' (Dickenson and Clarke 1972, 25). This provided an important counterpoint to the clarion calls for a policy-relevant geography (Berry 1972, 77) and for the development of a corpus of normative statements within the subject (Chisholm 1971a, 65): indeed, Prince (1971, 151) was convinced that so few geographers ever stopped 'to reassess the value of their contributions in other than strictly commercial terms' as to amount to 'a surrender of academic freedom on the part of those depending on research funds'. Although the cause of freedom was being invoked, however, it was not simply a case of demanding academic detachment, but rather of insisting on a prior explication of the terms on which commitment was to be made, *before deciding whether or not to accept them.*

There appeared to be two ways out of this moral dilemma. On the one hand it was possible to reiterate the case for autonomy, and to uphold what Chisholm (1969, 8) described as 'a strong and honourable tradition in the subject of teaching as the main *raison d'être* of university departments and of research as an essentially personal and private affair undertaken by teachers at least as much to enhance their personal career prospects as for other reasons'. Thus, twenty years after Dudley Stamp's (1951, 5) essay on applied geography had insisted that it is 'utterly wrong to confuse a scientific survey with a *policy* which may afterwards be developed in consequence of its findings', Robson (1971, 137) reaffirmed that there was

still a convincing case to be made for divorcing academic study from value judgements, and while it may be hard to argue that any academic study is ever likely to free itself completely from such values, this is not to suggest that geographers ought to be guided in their choice of research topics towards the value judgements themselves. It is one thing to point to the influence of implicit values on the derivation of theory: it is another to insist that the values themselves should be an object for study or indeed that the respective merits of different sets of values can somehow be resolved by academic review.

There was a fundamental difficulty in any attempt to secure such detachment, however, as Prince (1971, 152) pointed out: to restructure the domain of social science necessarily involved an intervention in the domain of public policy, since 'institutions practising discrimination through selection procedures, fostering conflict through competitive examinations [and] maintaining hierarchies through divisions between students and faculties [are] unable to generate free and independent thinking or to create an atmosphere in which all the world might be examined and criticised without fear or favour, yet with sympathy and respect'. To argue that the university should refrain from making any political stand on social issues was, therefore, as White (1972, 103) maintained, to 'ignore the possibility of its doing so – for many of its corporate actions such as admissions, property management and governance inevitably involve value judgements'.

To some extent, of course, the ideal of detached inquiry had in practice long been discarded. Indeed, Gouldner (1976, 180–1) speaks of it as historically specific and purely transitory:

It was because of a unique historical situation that, following the French Revolution, the unattached intelligentsia could take important initiatives and had much autonomy. This intelligentsia faced a discredited aristocracy and clergy on the one side, and on the other a publicly inexperienced, only precariously legitimate new bourgeoisie, who were separated from the direct exercise of violence or from the production of the culture and ideas that might legitimate them. The newly unattached intelligentsia could thus live in the interstices and unresolved antinomies between elites, between an old aristocracy that was historically outmoded and a new bourgeoisie that was historically immature. This unattached intelligentsia might then curry the favor and custom of the new bourgeois to his face, while sniggering at him in the coffee shops.

The course of events in other countries need not detain us; even if Gouldner's thesis merits elaboration or qualification in detail, as it probably does, it is, I suspect, firm enough in outline to sustain the subsequent closure of the interstices which the alternative solution accepted.

On the other hand, therefore, it was possible to recognize that, for various historical reasons, geography *had* become involved in the domain of policy-making, and that it was unlikely to be able to extricate itself in the foreseeable future. But if this meant that it could no longer claim a cloistered existence for itself, then it could at least afford to be more open about the status of the claims which it did make. 'What matters', said Blowers (1972, 291), 'is that the values upon which research is based are made explicit', so that 'we may unmask the false objectivity that covers much of our research and confuses our interpretation of findings'.

The two most fully developed responses to this demand for a new honesty within geography were those of Buttimer and King. Buttimer, couching her argument in terms of existentialism and phenomenology, concluded that an awareness of the 'basic values which permeate a person's mode of being-in-the-world' should 'help the scholar to reduce or eliminate the inconsistencies between thought and action', by allowing him 'to recognise the values and meanings which are implicit in his modes of construing reality, and the evaluative emphases of whatever model conceptualization of reality he uses' (Buttimer 1974, 39). King, building on this *value-aware* approach, advocated a *value-critical* one: one which would not only 'recognise the ethical content of all its analyses and freely admit the biases which are inherent in them', but which would also, Robson notwithstanding, make the values and goals themselves subject to analysis and debate (King 1976, 294 and 306). The most effective way of doing this, he believed, was through 'story-telling' (Rein 1976). This relies on the use of metaphors 'to bring two separate domains into cognitive and emotional relation by using language directly appropriate to the one as a lens for seeing the other; the implications, suggestions and supporting values entwined with the literal use of the metamorphical expression enable [us] to see a new subject matter in a new way' (Black 1962, 236–7). This explicitly creative engagement between two different frames of reference resembles the hermeneutic encounter, and Rein (1976, 87–8) treats immersion and part–whole relationships in terms close (but not identical) to those employed by Gadamer.

To summarize. The materials assembled so far suggest that geography may be passing through what Bartels (1973) identifies as a third 'wave of rationality'. The first, dominated by instrumental reason and the technical control of the environment, swept through the discipline in the 1950s and early 1960s. The second began to move soon afterwards, driven by the realization that the preceding claims for a scientific geography were grounded in a life-world which necessarily allowed a *plurality* of basic starting-points. The third, towering above the other two, will, if Bartels is correct, provide the energy for an 'historical–critical understanding of the relative positions of [these] various geographical perspectives'.

But this is still very much in the future since, he argues, it will eventually mean accepting that social science is 'tied to socially-determined preconceptions by a hermeneutic circle' (Bartels 1973, 33–4). *Nowhere in the preceding review has this possibility been taken seriously.* It is one thing to admit, with Smith (1977, 5) that 'radical' geography is the geography of the years of Viet Nam and Cambodia, of Northern Ireland, of Watergate, of economic collapse and environmental crisis, but it is another to establish the precise bonds which tie systems of ideas to structures of society. It may well be true, as Overton (1976, 87) suggests, that 'we are starting to examine and to understand the function in capitalist society of knowledge in general and geography in particular'; but we *are* only starting, and the belief that we can get very far without seeing whether the practical implications of geography are in some way predetermined is, I believe, fundamentally misplaced.

To say this is not, of course, to deny the importance of the *self-*questioning and the sense of *personal* responsibility which so many have now urged upon the profession. But simply to assert – for example – that 'the city needs not new structures but new men' and to locate 'the root cause of evil in the city' in 'privatistic iniquity not social iniquity' (Ley 1974, 68–71) is to close off the mediations between the two which would give the Christian doctrine the practical force which Ley wishes to claim for it. (It also proscribes any critical reflection on the doctrine itself, inasmuch as it occludes the relationships which exist between organized religion and social structure.) Equally, therefore, the metatheoretical considerations proposed by Bartels do not abandon the politics of illusion, the blind faith in individual autonomy, for a politics of despair, a resigned accommodation to the *status quo*. For his argument is that the hermeneutic circle can (and indeed must) provide the impetus for a fourth wave of rationality: 'the critical *formation* of science as a task of society and certainly above all as a task of the scientist himself' (Bartels 1973, 35). The reason for his emphasis on 'formation' can be clarified through a consideration of the conception of critique embodied in the programme of the Frankfurt School.

From Horkheimer to Habermas

The 'Frankfurt School' is, as Slater (1977, 26) says, a loose label applied retrospectively. But if the school never achieved any real coherence, let alone sustained an orthodoxy (Lenhardt 1976, 34), it was nevertheless distinctive enough to be widely regarded as one of the most significant intellectual movements to come out of Weimar Germany. Its members preferred to be identified by what they called 'the critical theory of society' rather than through any corporate affiliation, but this is still far from providing a more precise appellation inasmuch as it conceals what Connerton (1976, 15) calls their multiple allegiance.

Critique as they used it had at least three meanings. The first of these was derived from the Enlightenment. This has already been described as the age of reason, and critique in this sense was its 'essential activity', the subjection of all spheres of life – including, at last, the state itself – to a rational and public scrutiny which strengthened 'the long-established claims of reason as a means of knowing natural law' (Hawthorn 1976, 9). This is not to ignore the way in which the rhetoric of the Enlightenment 'was sedimented with a layer of class and elite interests' lying beneath its claims to be a universal public discourse (Gouldner 1976, 227), of which the Frankfurt School was profoundly aware, but rather to note a comparable commitment to critique as an activity of 'unveiling' or 'debunking' (Connerton 1976, 16–17).

The other two meanings of critique were more precise than this one, but they were closely related to it. They were derived from Kant and from Hegel.

The Kantian legacy was concerned with 'the proper problem of pure reason', and although it owed much to the earlier contributions of Hume and Rousseau it amounted to much more than a joint account. Thus while Kant agreed with Hume that experience was inscribed in knowledge he also maintained that the latter could not be directly grounded in the former: if this were the case, then 'our sense impressions would remain sense impressions and we should not be able to organise them into propositions'. Instead, therefore, these abstract concepts had to be located in the 'thinking subject' itself, the 'I' which actively constitutes the experienced world, and since reason was held to operate with *universal* categories then what Rousseau had described as this 'salutary organ of the will' could, in principle, disclose *universal* directives to action which, given their unconditional application to all men, would of necessity be morally binding (Hawthorn 1976, 33–4). Kant remained sceptical about the likelihood of such an historical disclosure being realized, particularly when he saw how his critique was taken up by those claiming to share it, let alone by those who openly rejected it, but nevertheless he continued to try to clarify the grounds on which it was at least *possible*. Critique in this sense, then, came to mean 'the rational reconstruction of the conditions which make language, cognition and action possible' (Connerton 1976, 18).

The Hegelian tradition, which developed partly as a reaction to Kant, was concerned with the dialectic of reason (see above, p. 109) and its birthplace was the *phenomenology of Mind*. Hegel's progression of thesis–antithesis–synthesis presupposed that the logical structure of the present carried within itself the logical structure of the future. This meant that if the world were moved by the Hegelian Spirit, critical reflection on the historically specific forms which it assumed would enable man to free himself from the illusion of their permanence. Critique in this sense was a reflection on and hence a transcendence of the evanescent structures which constrained human action, and as such necessarily entailed a conception of emancipation (Connerton 1976, 20).

All three meanings of critique can be found in the work of the Frankfurt School, but their transparent idealism was transformed through a deliberate encounter with Marx's materialism. The Institute's first Director, Carl Grünberg, who had left his post at the University of Vienna for a Chair of Economics and Social Science at Frankfurt, made it clear in his inaugural lecture that 'the method taught as the key to solving our problems will be the Marxist method' and although, as Jay (1973, 12) observes, his understanding of what precisely this involved was later to be questioned – and even abandoned – by the younger members of the Institute, it was nevertheless to prove a lasting declaration of principle. Only four years after his appointment Grünberg suffered a stroke and in 1931, at the age of thirty-five, Max Horkheimer assumed the Directorship. He moved the Institute away from the somewhat mechanical – 'unimaginative' (Jay 1973, 12) – concerns of his predecessor and towards what, very early on, he described as the 'relation between the various areas of material and mental culture' (Slater 1977, 14).

This is in fact the very intellectual topography which was mapped, in a rough and ready way, into Bartels's discussion of rationality in geography. Its contours can be fixed more precisely through an examination of Horkheimer's programmatic essay on 'Traditional and Critical Theory', published in 1937. (This was, incidentally, written in exile: as Jay (1973, 29) says, once Hitler came to power in 1933 'the future of an avowedly Marxist organization, staffed almost exclusively by men of Jewish descent – at least by Nazi standards – was obviously bleak' and the Institute was closed within a matter of months for 'tendencies hostile to the state'.)

In his essay Horkheimer characterized traditional theory as a scientific activity which claims a determined autonomy or detachment for itself: 'the construction and validation of theory are activities carried on without regard for the purposes which it will serve' (Lewis and Melville 1977, in press), with the result that science is elevated above any sectional interest. It follows from this that 'the scholarly specialist "as" scientist regards social reality and its products as extrinsic to him, and "as" citizen exercises his interest in them through political articles, membership in political parties or social service organizations, and participation in elections. *But he does not unify these two activities*' (Horkheimer 1972, 209–10;

italics added). The virginity of science thus depends on whole legions of Dukes of York marching up to the top of the hill and marching down again, moving from (and distinguishing between) their responsibilities as scholars above and as citizens below. It is, of course, this sort of fairy-tale image which has been projected on to the debate about 'relevance' in geography from its inception but, to continue the analogy, it is unlikely – to say the least – that the maidenhood of science can be preserved in the midst of military manoeuvres. Horkheimer explained it like this:

> The traditional idea of theory is based on scientific activity as carried on within the division of labour at a particular stage in the latter's development. It corresponds to the activity of the scholar which takes place alongside all the other activities of a society but in no immediately clear connection with them. In this view of theory, therefore, the real social function of science is not made manifest; it speaks not of what theory means in human life, but only of what it means in the isolated sphere in which for historical reasons it comes into existence. Yet as a matter of fact the life of society is the result of all the work done in the various sectors of production. Even if therefore the division of labour in the capitalist system functions but poorly, its branches, including science, do not become for that reason self-sufficient and independent. They are particular instances of the way in which society comes to grips with nature and maintains its own inherited form. They are moments in the social process of production, even if they be almost or entirely unproductive in the narrower sense. Neither the structures of industrial and agrarian production nor the separation of the so-called guiding and executory functions, services, and works, or of intellectual and manual operations are eternal or natural states of affairs. They emerge rather from the mode of production practised in particular forms of society [Horkheimer 1972, 197].

Horkheimer concluded from this that so-called 'common sense' is regarded as a court of appeal 'for which there are no mysteries' not because it somehow corresponds to a domain of immanent truth but because 'the world of objects to be judged is in large measure produced by an activity that is itself determined by the very ideas which help the individual to recognize that world and to grasp it conceptually'. And this uniformity is achieved not through an anonymous Kantian subject, nor through an absolute Hegelian Spirit, but through the structuration of the mode of production. The scientific activity of any society, according to Horkheimer, is but 'a moment in the continuous transformation and development of the material foundations of that society' (Horkheimer 1972, 194) and even if, as Lewis and Melville (1977, in press) suggest, he is using the concept in a restricted sense here, it is nevertheless this relation to the mode of production or, better, the social formation, which allows and obliges us to speak of the *formation* of science.

It must follow from this, however, that what Horkheimer (1972, 206) called 'critical activity' also emerges from the social structure; and yet at the same time he insisted that 'its purpose is not, either in its conscious intention or in its objective significance, the better functioning of any element in the structure'. For this to be possible, Horkheimer had to claim that critical theorists 'interpret the economic categories of work, value and productivity exactly as they are interpreted in the existing order, and they regard any other interpretation as pure idealism. But at the same time they consider it rank dishonesty simply to accept the interpretation; *the critical acceptance of the categories which rule social life contains simultaneously their condemnation*' (italics added). This notion of science as examined discourse ought, by now, to be a familiar one, but its importance is not simply in its 'explicit recognition of the connection of knowledge and interest' (Bernstein 1976, 180), and neither does it reside wholly in its dialectical, transcendent motion, its vision of 'a transcended, reconstructed world' (Agger 1977, 13). It also obliges the critical theorist to speak *directly* to those involved, to use and move beyond their *own* categories and cognitions: and in fact Horkheimer (in Connerton 1976, 224) was convinced that critical theory could become 'an essential element in the historical effort to create a world which satisfies the needs and powers of men' *only* to the extent that it was able to engage its subjects in this way.

Even so, it has to be admitted that in real terms the project met with only limited success. There were several reasons for this, not least among them being a spectacular failure to speak in a language which *could* be understood outside a narrow circle of *cognoscenti*. But over and above this Bernstein (1976, 184) believes that the really major weakness of the Frankfurt School was its lack of 'a sustained argument that move[d] from traditional to critical theory. Such an argument requires showing how the conflicts and contradictions inherent in traditional theory force us to move beyond it. Otherwise we face an impasse where one is in effect being told, "Here I stand and there you stand" '. And in many ways this has been the most serious deficiency of the radical resurgence in geography as well. It has allowed some geographers to claim to be standing here on some occasions and there on others: so, for example, Chisholm (1975, 175) contends that it is a mistake 'to suppose that because one paradigm (the "scientific" [positivist] paradigm) is inadequate to solve *all* our problems it is *altogether* useless'. To some, of course, even this admission might appear as a strategic victory; but to abandon the struggle as a result of it would be to deflect the full force of the Frankfurt School's critique. While its members certainly registered their objections to closed philosophical systems, and indeed developed their ideas through close interactions with other schools of thought (Jay 1973, 41), they just as clearly resisted any eclecticism which threatened to attenuate their critical intentions. In other words, there is a danger that attempts to assimilate critical insights into the corpus of traditional geography will leave its foundations undisturbed and its primary allegiances unchallenged: negation can follow hard on the heels of recognition, no matter how elaborate the ceremony, and the possibility ought not to be taken lightly.

But having said that, arguments like Chisholm's can be given a radically new edge by turning to the philosophical anthropology of Jürgen Habermas. To connect the two of them in this way will perhaps seem mischievous, but my reason for doing so is simply that in recognizing the limited applicability of what, returning to my earlier usage, I will now call the *empirical–analytic* sciences and, by extension, presumably upholding the (co-equal?) claims of the *historical–hermeneutic* sciences, Chisholm must, if Habermas's discussion of knowledge-constitutive interests is correct, accept the imperatives of the *critical* sciences, which involve the conjunction of the other two (see above, pp. 69–70). To say this is evidently to initiate an ambitious project, and in trying to provide such a grounding for critical theory – the trajectory missing from Horkheimer's contribution – Habermas moves outside the confines of the Frankfurt School. It is impossible to pursue him very closely here, particularly since his project is still provisional and programmatic, but the concept of central importance to our present discussion is clearly that of 'interest'.

I have already said that Habermas recognizes an interest in *technical control* in the empirical–analytical sciences, an interest in *mutual understanding* in the historical–hermeneutic sciences (the 'practical' interest), and an interest in *emancipation* in the critical sciences. The epistemological status of these three knowledge-constitutive interests is, as Bernstein (1976, 192) observes, one of the most problematic features of Habermas's work, and this is certainly not the place to attempt a clarification of the 'quasi-transcendental' status which he originally ascribed to them. He himself now admits that the formula was 'a product of an embarrassment which points to more problems than it solves' (Habermas 1974, 14). It arose out of the need to find a way of identifying the interests which would locate them neither in a transcendental subjectivity (which would put them beyond the reach of history) nor in a contingent empiricism (which would allow them to be determined by individual intentions). Instead, they were to represent 'general cognitive strategies of the action-related organization of experience' (Habermas 1974, 21). The consequences of this are of more immediate significance to us, since the claim I have made for the imperatives of the critical sciences rests on an equivalence between the three interests, and their connection with domains of action will enable us to sort out what is involved much more clearly.

Habermas ties the first two interests to *production* and *interaction* respectively, and even if it is difficult to prise the one from the other as cleanly as perhaps he would wish it is at least plausible to regard them as the primary levels of social action which he requires them to be, and possible to see how they might determine different objects of study and establish different criteria for making valid statements about them. But he ties the third interest to *power* which, although lending renewed force to Hägerstrand's vision of the core of a new human geography, appears to introduce a basic asymmetry into the scheme. Whereas the first two seem to specify only the form which knowledge must take, the emancipatory interest seems to dictate its content as well (Bernstein 1976, 209). Habermas needs to show, therefore, that the specification of an emancipatory interest is not an arbitrary, normative impulse on his part but one which, just like the technical and the practical, *is necessarily grounded in the existing structure of social action*. He tries to establish the equivalence of all three interests in this way by making what has come to be known as the 'linguistic turn', into a theory of communicative competence.

His strategy is to maintain that any successful form of communication depends on a mutual presupposition that four different 'validity claims' (*Geltungsansprüche*) are met. These are:

(1)   that the statement is comprehensible;

(2) that the propositions in it are true;
(3) that the speaker is in a position to make the statement;
(4) that the speaker means what he says.

This 'background consensus' can be fundamentally disrupted by calling any one of these claims into question, and there is then no obvious way of re-establishing it, because any attempt to justify the disputed validity claim through renewed argument must necessarily involve and rely upon the consensus which has just been disturbed. The only way out of this – and Habermas insists that it is a way which we all take, whether we realize it or not – is to assume that the consensus is tied to an *ideal speech situation*, one which is so free from internal and external constraints that the force of the better argument must necessarily prevail (McCarthy 1973, 153). That such a guarantee of rationality is intrinsically empty is beside the point. All Habermas is seeking to show is that, as he put it in his inaugural lecture, 'our first sentence expresses unequivocally the *intention* of universal and unconstrained consensus' (italics added): that the ideal speech situation must be anticipated in any 'language-game'. It is then a small step to conclude that this must at the same time presuppose an ideal form of social life.

Habermas's analysis, which is much more intricate than this simple sketch can suggest, raises many questions, particularly when we descend from its metatheoretical heights to ask whether it is possible to specify the relationships between critical science and the ideal form of social life in any more concrete form. Habermas's own answer is extremely cautious, and in fact he seemed so evasive and equivocal in the face of the struggles of the late 1960s that he earned the bitter condemnation of the student Left. The reason for their impatience is not hard to see, and in what follows – and bearing in mind that there are many other facets of his project that deserve comment – I will concentrate on Habermas's view of the relationship between theory and practice, and try to relate it more specifically to the conduct of a committed geography.

To anticipate the argument: the theory of communicative competence may provide some sort of warrant for the emancipatory interest, but it cannot legitimize the various courses of action which promise to realize the ideal form of social life. In particular, Habermas (1974, 32) is adamant that 'there can be no meaningful theory which per se, and regardless of the circumstances, obligates one to militancy'. To claim that positions like Chisholm's involve accepting the imperatives of critical science, therefore, is to claim that they involve accepting a conception of social science as critique – no more. The distinctions between this view and the traditional one should not be minimized, but if it is hardly surprising that such a conception should appear excessive to the traditionalists (who regard it as ideologically motivated), then it is no more surprising that it should appear insufficient to the activists (who regard it as politically bankrupt). It will be clear by now that the first of these reactions is misplaced. The three interests are not barriers to the objectivity of knowledge (as the traditional belief in 'pure theory' presupposes), but the very *conditions* of objective knowledge, and this is as true of the emancipatory interest as it is of the technical and the practical. We can turn, therefore, to the second reaction.

Theory and practice in geography

Wellmer (1976, 258) describes Habermas's conception of social theory as a critique which

penetrates beneath the surface grammar of a 'language-game' to un-cover the quasi-natural forces embodied in its depth-grammatical relationships and rules; by spelling them out it wants to break their spell. Its internal *telos* is to enhance the autonomy of individuals and to abolish social domination and repression; it aims at communication free of domination. Such a critical theory, consequently, can become 'practical' in a genuine sense only by initiating processes of self-reflection – a self-reflection which would be the first step on the road toward practical emancipation.

The extent to which critical theorems can successfully initiate (and be sustained by) these processes of self-reflection is dependent on the extent to which empirical–analytic and historical–hermeneutic forms of science can effectively be brought into conjunction.

In describing the project like this Wellmer is relying on Haber-mas's debt to structural linguistics, and especially on the connection between the theory of communicative competence and Chomsky's model of linguistic competence. This inevitably distances Habermas from the narrowly positivist interpretations of the empirical–analytic sciences which often seem to be present, particularly in his early work, in so far as it dissociates him from the traditionalists' closure of inquiry around the phenomenal level. At the same time, however, it invites the suspicion of activists, who reject purely linguistic solu-

tions as counter-revolutionary stratagems (and are right to do so).
But it is important to realize that Habermas is in fact seeking to
'concretize' the conception of critique which he advocates: in other
words, to turn the advances made in linguistic philosophy *back*
upon themselves, and so *transcend* the categories of existing
theoretical structures (Frankel 1974). In this sense his critical intention is the same as Horkheimer's (although obviously given a different expression), and for substantially the same reason: the need
to engage the subjects of critical theory in its formation.

The first attempts to end modern geography's cultural isolation
through deliberate engagements of this sort were sporadic and short-
lived, and it would be idle to pretend that they owed anything to
Habermas. Most of them took the form of 'geographical expeditions'
whose rationales emerged gradually in the course of encounters with
the local, typically urban, community. (I am excluding earlier ventures, like Mackinder's famous Extension lectures or the even
earlier promulgation of geography through the classes of the
Mechanics Institutes, since they were all founded on a traditional
pedagogic model and can hardly be described as critical in either
conception or execution.)

The Detroit Geographical Expedition, for example, began to
crystallize during the summer of 1969, when its defining characteristics 'moved from the vague thoughts and desires of the individuals who founded the organization to the point where many of
these thoughts were actually articulated on paper and some put
into practice' (Horvarth 1971, 74). The University of Michigan
eventually agreed to validate an extension course on the geographical aspects of urban planning. The idea, says Horvarth, was 'that
any black person could walk in off the street, take 45 hours of
university credit courses free of charge, and then transfer with
sophomore status to any Michigan university if he could do C or
better work'. In addition, and in particular, the programme was to
be run 'by the people from the community it serves, who act not
as an "advisory board" but as the total administration'. As the first
course ended so did Ann Arbor's connection with the Expedition,
and Michigan State University stepped in to offer three similar
courses. This time the problem was not the intransigence of university bureaucracy but the content of the course itself. 'A lot of effort
had to be spent in shoring up the morale of students, who had real
difficulties just showing up even to free classes – some came hungry,
others couldn't afford bus fares, one student had been living in a

F

car for five weeks. Learning how to make a clean line, lay a zip-a-tone pattern, or design a map with the right combination of point, area and line symbols did not seem to be critical knowledge to members of a survival culture.' Fortunately the Expedition was then asked to provide assistance for the educational redistricting of the city of Detroit, and this at last 'made sense. The next three weeks both saved and came to define the potential of the Expedition' (Horvarth 1971, 77). The report which resulted showed that four of the eight proposals put forward by the Board of Education were illegal, and outlined an alternative scheme which was subsequently adopted by the Northwest Community Organization. Horvarth (1971, 79) regards this as an indication 'that the principles upon which the Expedition rested were not only sound but powerful'.

These principles were incorporated into later expeditions, and when the practical lessons which they contained were translated into theoretical terms the language was Friere's. Thus Campbell (1974, 103) argued that the presence of what he called 'advocate geographers' ought to allow members of the local community 'to become problematizers of their situations and to become active creators of their environment'. 'It is not enough', he said, 'to allow the community to define *some* of the problems and to carry through *some* of the solutions; the whole enterprise must be reciprocal and committed to genuine partnership.' This draws on Friere's (1972) theory of dialogical action, and the parallels between this and Habermas's project need to be spelled out.

Dialogical action is committed to liberation (emancipation) in such a total sense that it has to struggle against being subverted by strategies which, appearances to the contrary notwithstanding, are committed to domination. It is for much the same reason that Habermas rejects any prior guarantees for critical theory:

Decisions for the political struggle cannot at the outset be justified theoretically and then carried out organizationally. The sole possible justification at this level is consensus, aimed at in practical discourse, among the participants, who, in the consciousness of their common interests and their knowledge of the circumstances, of the predictable consequences and secondary consequences, are the only ones who can know what risks they are willing to undergo, and with what expectations. There can be no theory which at the outset can assure a world-historical mission in return for the potential sacrifices [Habermas 1974, 33].

To deny the hermeneutic encounter, then, would necessarily circumscribe the effectivity of critical theory because it would reduce its strategies to narrowly technical ones of manipulation and control. Similarly, Friere (1972, 108–9) points out that strategies like this are predicated on a 'necessity for conquest', through which the conqueror 'imposes his own contours on the vanquished, who internalise this shape and become ambiguous beings "housing" another'. The most obvious way in which this is done, he suggests, is by 'precluding any presentation of the world as a problem and showing it rather as a fixed entity, as something given – something to which men, as mere spectators, must adapt'. A critical science realizes its self-image in its response to this, of course, but it must ensure that its problematization does not substitute a revolutionary teleology to which men, as mere instruments, must be bent. 'Revolutionary praxis is a unity', warns Friere (1972, 97), 'and the leaders cannot treat the oppressed as their possession': if they do, they betray the hope of emancipation and restore the panoply of domination.

But this is not the only way in which self-reflection can be compromised. A critical science cannot afford to let its problematization be devalued through relying on 'a *focalized* view of problems rather than on seeing them as dimensions of a *totality*' (Friere 1972, 11). Again, Habermas makes much the same point. This sort of problem-orientation has found its way into most social sciences in recent years, and it carries with it an open invitation to co-optation. The style of research which defines a concern with misery, hunger and oppression as a concern with a world somehow more 'real' than the kind of experiences which have traditionally constituted the domain of academic inquiry is only its most strident expression. *It has no place in a genuinely critical science* in so far as it directs attention towards focalized solutions which obscure and ultimately reinforce the basic structures of society. It has this effect because the distinction it makes between 'more real' and 'less real' worlds is one which readily assumes that assaults on the one need not (and even must not) penetrate the other; that the relationship between them is sufficiently contingent for the warts and blemishes to be removed by a skin-deep surgery which leaves the complexion untouched. As Colenutt (1976, 85) puts it, as long as someone is working on the 'more real' world, as long as our understanding of the totality remains fragmented, then the basic structures of society and the forms of knowledge they legitimize remain unexamined. In the end they are

built into social engineering designs as fundamentally sound plat-
forms from which to launch individual reforms, and the much-
vaunted problem-orientated geography becomes just another expression of 'the reforming ideology of contemporary capitalism'.

Harvey (1973, 152) suggests that there are two important revolutionary tasks in theory formation, and both of them can be connected to these preliminary considerations. One is *negation*, which
involves taking the compromised formulation 'and exposing it for
what it is', and the other is *reformulation*, which involves transcending the existing theories 'and using them to identify real choices
immanent in the present'. The success of these tasks, I suggest,
depends on the ability of a theory of communicative competence to
explain how 'the structure of communication allows some issues to
be seen and spoken, while diverting attention from still others'
(Gouldner 1976, 146). Habermas (1974, 12) has in fact recognized
this, saying that he wants to develop the theory of communicative
competence in historically specific terms and to construct a theory
of 'systematically distorted communication' consonant with the
'fundamental assumptions of historical materialism'.

In outline, his argument rests on the claim that normative impulses
are capable of rational justification. In taking up a practical discourse (one through which validity claims are disputed) we un-
avoidably suppose an ideal speech situation. 'If under these conditions a consensus about the recommendation to accept a norm
arises argumentatively, that is, on the basis of hypothetically proposed, alternative justifications, then this consensus expresses a
"rational will".' We are entitled to regard it as rational, Habermas
says, because the formal properties of practical discourse 'guarantee
that a consensus can only arise through appropriately interpreted,
generalizable interests, by which I mean needs *that can be com-
municatively shared*' (Habermas 1976a, 108). It must follow from
this that norms which do not regulate generalizable interests are
based on domination rather than argumentation so that, by extension, critical theory has to be capable of exposing them 'for what
they are'. Habermas's point is that contemporary capitalism is
legitimized through the exercise of precisely this kind of 'normative
power', and that it is expressed through formulations

that have the double function of proving that the validity claims of
norm systems are legitimate and of avoiding thematization and testing
of discursive-validity claims. The specific achievement of such ideo-

logies consists in the inconspicuous manner in which communication is systematically limited. A social theory critical of ideology can, therefore, identify the normative power built into the institutional system of a society only if it starts from the model of the suppression of generalizable interests [Habermas 1976a, 112–13].

The start is admittedly still only a rudimentary one, but it is nevertheless sufficiently complete for two conclusions to be drawn from it. First, if critical theory is not to surrender its political relevance it 'needs an emphasis on the institutions connecting the state and language' (Gouldner 1976, 149). It is simply not enough to construct metatheoretical frameworks, important though they are; critical theory has to provide an understanding of the structure of *specific* theoretical systems and of *specific* social mediations. In so far as it is possible to isolate one discipline's contribution to what is effectively a programme of ideology-critique, we can clarify the role of geography. Gouldner (1976, 152) suggests that 'the central semiotic effort of modern politics is the capturing and evocation of a symbolism of freedom and/or equality'. If this is accepted, then geography's part in the critique of such an effort is likely to consist of (a) a critique of the concepts through which the discipline has sustained its image of the world; and (b) a critique of the processes through which the social formation has sustained its relations of production. The difficulty of speaking within the existing fragmentation of inquiry is at once revealed, however, because neither of these tasks can be accomplished until the traditional barriers between disciplines are removed. (Not that this is the only condition of success.) Most of the concepts used in geography are derived from other disciplines, and a focus on narrowly 'geographical' processes – even supposing that we could agree on what they were – could only partially disclose the structure of the totality. To draw back from the barriers, therefore, is to disengage from the struggle. This assumes all the more force once the close connections between the two tasks are fully recognized. Marx's critique of political economy put it succinctly: only by transcending the theoretical domain in which 'rule freedom, equality, property and Bentham' is it possible to show how the process of exchange described there is but 'the phenomenon of a process taking place behind it' and to reconstruct the relations of production which constitute the practical domain. To repeat, then, 'there are no short-cuts if the potential of geography is to be realized *within an integrated framework.* It requires a thorough critique of existing geography, which is at the same time a critique of the geography of objective reality' (Anderson 1973, 5). And since neither of these are likely to emerge unscathed it is perhaps not surprising that Friere (1972, 99) should proclaim that 'critical reflection is also action'.

To some, no doubt, this will seem a grotesque prostitution of what action ought to entail, and this brings us abruptly to the second conclusion. If we stay within the programme of ideology-critique set out by Habermas, we can say that geography's major response so far (and there have been others) has been (a) to develop a sustained critique of the concept of rent and, through this, (b) to show how residential differentiation acts as 'an integrating mediating influence in the processes whereby class relationships and social differentiations are produced and sustained' (Harvey 1975, 368). Both of these efforts were initiated by David Harvey and, to speak with Habermas, they expose the manipulation of the background consensus and the suppression of generalizable interests. The way in which the first of these projects resists the systematic distortion of communication ought to be obvious; the effect of the second is to expose the way in which residential differentiation fragments class consciousness into a 'community consciousness' which, if not exactly a culture of collaboration, is at least a culture of consolation which is not readily transformed into effective praxis (Stedman Jones 1974; Harvey 1975).

Both of these projects, then, challenge normative power, and in so far as this is exercised in the name of the corporate state it is clear that efforts like them must attack – at least subvert – geography's enlistment in its service. Indeed, Gouldner (1976, 150) argues that in so far as such actions can be connected to Habermas's model of the suppression of generalizable interests they must imply the *capture* of the state. Hence, and to particularize, Harvey (1974a, 24) insists that 'the moral obligation of the geographer, *qua* geographer, is to confront the tension between the humanistic tradition and the pervasive needs of the corporate state directly, to raise our consciousness of the contradiction and thereby to learn how to exploit the contradiction within the corporate state structure itself'. But Gouldner wants to call attention to a contradiction which he claims exists *within praxis itself*: 'the means to bring about the communicative competence that Habermas requires for rational discourse presuppose precisely the centralization and strengthening of that state apparatus which increasingly tends to stifle rather than facilitate the universalization of the rational, uninhibited discourse necessary for any democratic society' (Piccone 1976, 173).

This conclusion is palpably uncomfortable. Some will no doubt immediately dismiss it as bourgeois sentimentality – those who embrace what to Gouldner (1976, 151–2) is 'a fading, unreflective Leninism that views politics as a kind of career; as the sublimation and routinization of heroism' in which a revolutionary vanguard applies 'the steel forceps to history to extract the reluctant revolution'. Still others will immediately dismiss it as a demonstration of the fundamental inadequacy of Habermas's project – those who are unwilling to pursue the consequences of their commitments, and who prefer to keep socialism soft in their minds rather than allow it hard into their lives. Habermas (and Gouldner) may well be wrong, their arguments flawed and their concerns misplaced, but the only authentic response which can hold out the hope of a different solution with any integrity is one which takes their conclusion seriously: which explores the theory from which it derives and the practice to which it appeals.

Geography has always lived in the shadows between these two domains, hiding from open encounters with the theories which direct it or the practices which sustain it. These pervasive timidities will not easily be overcome, and to expect geography to welcome intruders like this, from within or without, is to expect too much. But, fortunately, this does not matter: like any form of inquiry, geography is balanced on what Griffiths calls 'the essential connective imperative'; no matter how hard it tries it will not be able to snap it. And in the end it will be unable to resist its exposure. Its outward detachment will be overcome and its inner commitment revealed.

Conclusion:
A place among the social sciences

With his hands clasped in his lap he let his eyes swim in the wideness of the sea, his gaze lose focus, blur, and grow vague in the misty immensity of space. His love of the ocean had profound sources: the hard-worked artist's longing for rest, his yearning to seek refuge from the thronging manifold shapes of his fancy in the bosom of the simple and vast; and another yearning, opposed to his art and perhaps for that very reason a lure, for the unorganised, the immeasurable, the eternal – in short, for nothingness. He whose preoccupation is with excellence longs fervently to find rest in perfection; and is not nothingness a form of perfection? As he sat there dreaming thus, deep, deep, into the void, suddenly the margin line of the shore was cut by a human form.

Thomas Mann: *Death in Venice*

The specific conclusions of this work are contained in each of the preceding chapters, and here I want to confine myself to a short summary of some of the more general conclusions which have threaded their way through each separate discussion, and to indicate some of the questions which they raise. This will involve moving beyond the text, as it were, and in doing so it will become clear that in certain strategic senses the problematic of the preceding chapters is *itself* unsatisfactory.

If we choose to retain geography's commitment to positivism (I speak in the collective advisedly: the room for individual autonomy within an institutionalized discourse is narrow), it should now be reasonably clear that this will involve our obscuring two major and inescapable issues of enormous practical significance. The first of these is the relationship between science and society. If a serious deficiency of Althusser's work was its failure to provide a coherent location for science in relation to the social formation, it at least took the task seriously and recognized its primary importance; this also provided the starting-point for the phenomenological critique, which sought to interrogate the connections between the conduct of a scientific discourse and the social conditions which make its particular processes and forms both possible and acceptable; and in his construction of critical theory Habermas rejected the possibility of an autonomous science altogether and insisted that conceptions of science are given in determinate social practices. These relations must be occluded in any geography which devotes itself purely to the extension of technical control, and their problematization must remain outside its grasp. It would of course be foolish to suggest that the technical interest ought to be resisted – it cannot be – but it *is* surely vital to resist its realization through any form of knowledge which conceives of and reproduces its conditions of existence in terms such as to destroy systematically the conditions of existence of other forms of knowledge.

And it is no answer to say, with whatever justification, that geography has never been nor can it ever hope to be a wholly positivist enterprise: that, irrespective of whether the aims of positivism are theoretically realizable on some high-level terrain, the day-to-day business of teaching and research is conducted on a low-level terrain whose blurred outlines cannot be fixed by the sharp contours of the philosopher's categorizations. This sort of inchoate eclecticism, however comforting it might appear, is no solution because – and this is the second issue – all conceptions of critical science require an effort of their practioners which obliges them to articulate the bases of (the interests in) their own practical activities. Bachelard and Althusser spoke of epistemological breaks; Husserl invoked a transcendental *epoché*; and Habermas warned of the casual co-optation of ostensibly critical formulations. That there is an urgent need for the problematization of discourse, for the exposure of connections between forms of knowledge and forms of existence concealed in and ultimately by social practice, is surely undeniable, and it is at least plausible to see why this must involve a break of some kind with the categories which we typically accept without demur. If we speak about residential differentiation, for example, we need to clarify the status of concepts like 'class' and 'rent' and to recognize their essentially and intrinsically political connotations. They cannot be used 'innocently'; interests are submerged in and are made to surface through them. But it is here, I think, that I am now at odds with some of the arguments contained in the preceding chapters. In opposing science to ideology, examined discourse to unexamined discourse, I suspect that I have risked (at

G

best risked) resuscitating the privileged status of what conventionally passes for science and even tried to free its propositions from
what are (equally conventionally) taken to be the ideological resi-
dues which impregnate the conduct of practical life. But the effort,
one is tempted to say the battle, required by a genuinely critical
geography cannot be understood like this at all; it must, at bottom,
require and enact a political decision, and the notion of somehow
distilling and purifying discourse from its commitments must be
strenuously resisted, the possibility denied.

The reasons behind this weak conception of what a critical geography entails strike at the very heart of the present text. They are
the product of striving towards a uniquely *epistemological* discourse,
that is, one which represents knowledge in terms of a correspondence between a domain of concepts and a domain of objects and
which seeks to specify – here, as and through the examination of
discourse alone – the conditions under which the two domains are
'really' coincident. This has resulted in a basic unevenness in the
text, inasmuch as the conflation between demonstration and assertion which it allows has meant that whereas I think I have managed
to demonstrate the inadequacies and the consequences of a traditional geography, I am conscious of – in the main, anyway – having
simply asserted the superiority of a critical one. This might be
excused, though certainly not defended, by emphasizing the provisional status of the work and, in any case, the previous chapter
has made some effort to identify the practical grounding of the
critical model and to spell out its political dimensions. But the fact
remains that, in effect, I have as yet failed to apply to the critical
model the same order of interrogation that I deployed against the
traditional model and, in particular, I have said very little about the
nature of the emancipatory interest which provides the touchstone
of the critique and the very foundation of the critical model itself.

These tasks are clearly urgent ones, and I have started to tackle
them in a subsequent volume (*Studies in Regional Geography*). The
issues are not confined to any one subject, of course, but taken
together they do at least testify that geography must be a much
more difficult enterprise than many of us have been prepared to
admit; and, I now want to suggest, it may even be one of the most
difficult of all. This is not, definitely *not*, because geography has to
play some kind of lone bridging role between the natural and the
social sciences. Claims like this have been advanced many times
in the past, as we all know, but more often as pious hopes or rueful

excuses than as serious propositions. These frequently turn on the belief that geography will *eventually* provide the grand synthesis, the contribution of substance, but *at present* it hasn't got very far because the natural and the social sciences keep pulling it in different directions. Presumptions of this kind have been grounded in either a series of misleading theoretical analogies which project from one domain on to the other (of which systems theory is only the most recent) or a casual empiricism which throws 'Man' into 'Nature' and lets him get on with it. Neither of these can be accorded scientific status, I shall say, because neither of them incorporate any systematic examination of discourse and *hence* of practical life into their projects. I have already said that this is not a sufficient criterion of what constitutes science, but it *is* enough to disqualify both of these attempts. What is more, of course, geography can hardly claim the dialectic between man and nature as its exclusive object of study: any properly constituted social science must address questions of structuration.

It is therefore high time to abandon the pretence of a separate existence for geography, one which has reduced it to a repository of low-level propositions which are at best self-evident and at worst simply false. The only exceptions turn out to be those which have been left on the intellectual trains of other disciplines: it is hardly surprising that geography should see itself as an umbrella subject.

What does make geography so difficult, it seems to me, is not these definitional problems at all, but rather its attempt to operate within specifically regional contexts. Ever since regional geography was declared to be dead – most fervently by those who had never been much good at it anyway – geographers, to their credit, have kept trying to revivify it in one form or another. (Even David Harvey's *Explanation in Geography* closes with a plea for some sort of regional geography.) This is a vital task: objections to the uncomfortable pinhead perch of neo-classical economics are familiar enough, but they also apply to the rest of political economy and social science. We need to know about the constitution of *regional* social formations, of *regional* articulations and *regional* transformations. To many, no doubt, this will seem obvious: it's not difficult to point either to the warrant provided by geography's long-standing commitment to places and the people that live in them or to the regional structures which persist in contemporary space-economies. But claims like this are really little more than special pleading – unsupported assertions which fail to substantiate the case for regional interventions in either intellectual or political terms. Again, I have treated this in more detail elsewhere (*Studies in Regional Geography*), but the nub of the argument is that spatial structures are implicated in social structures and each has to be theorized with the other. The result of this, if we still have to speak within institutional categories, is a doubly human geography: human in the sense that it recognizes that its concepts are specifically human constructions, rooted in specific social formations, and capable of – demanding of – continual examination and criticism; and human in the sense that it restores human beings to their own worlds and enables them to take part in the collective transformation of their own human geographies.

Guide to further reading

The listings which follow are confined to general texts, since the relevant works of more detailed focus ought to be clear from the preceding discussion and are cited in full in the Bibliography.

Probably the most compelling way to pursue the issues raised in these pages is to move with David Harvey from the positivism of his *Explanation in Geography* (London, 1969) through to the gradual and more critical reformulations of *Social Justice and the City* (London, 1973). Other important reflections are contained in the volume of essays edited by Richard Chorley under the title *Directions in Geography* (London, 1973). One of the most sustained and invigorating philosophical examinations of modern geography is Gunnar Olsson's *Birds In Egg* (East Lansing, 1975): a brilliantly flawed presentation which deserves the closest study.

An excellent elementary introduction to social theory is Russell Keat and John Urry's *Social Theory as Science* (London, 1975); more advanced, and important contributions in their own right, are Richard Bernstein's *The Restructuring of Social and Political Theory* (Oxford, 1976) and Anthony Giddens's *New Rules of Sociological Method* (London, 1976). Most difficult of all, but well worth the effort, is Barry Hindess's *Philosophy and Methodology in the Social Sciences* (Hassocks, 1977).

The most accessible historical survey of social theory, with many provocative insights on the way, is Geoffrey Hawthorn's *Enlightenment and Despair: a history of sociology* (Cambridge, 1976), and of economic theory Maurice Dobb's *Theories of Value and Distribution since Adam Smith: Ideology and Economic Theory* (Cambridge, 1973).

One of the most detailed examinations of neo-classical economics is Martin Hollis and Edward Nell's *Rational Economic Man: a philosophical critique of neo-classical economics* (Cambridge, 1975); the post-Keynesian (neo-Ricardian) alternative is set out by Joan Robinson and John Eatwell in *An Introduction to Modern*

Economics (London, 1973) and by J. A. Kregel in *The Reconstruction of Political Economy: an introduction to post-Keynesian economics* (London, 1973). Two excellent accounts of the Marxian critique are provided by Meghnad Desai's *Marxian Economics* (London, 1973) and M. C. Howard and J. E. King's *The Political Economy of Marx* (London, 1975).

Many of the most important contributions, of course, appear in journals and occasional publications, and with changes of editorial policy or emphasis it becomes difficult to indicate the key locations for critical essays; even so, major arguments have appeared in the *Annals of the Association of American Geographers, Antipode: A Journal of Radical Geography,* the *Canadian Geographer, L'Espace géographique, Espaces et sociétés* and the *Geographical Review.*

Bibliography

ACKERMANN, E. (1963), 'Where is a research frontier?', *Annals of the Association of American Geographers*, vol. 53, pp. 429–40

ADORNO, T., ALBERT, H., DAHRENDORF, R., HABERMAS, J., PILOT, H. and POPPER, K. R. (1976), *The Positivist Dispute in German Sociology*, London: Heinemann

AGGER, B. (1977), 'On happiness and the damaged life', in J. O'Neill (ed.), *On Critical Theory*, London: Heinemann, pp. 12–33

ALBROW, M. (1974), 'Dialectical and categorical paradigms of a science of society', *Sociological Review*, vol. 22, pp. 183–201

ALONSO, W. (1964), *Location and Land Use: Toward a General Theory of Land Rent*, Cambridge, Mass.: Harvard University Press

ALTHUSSER, L. (1969), *For Marx*, Harmondsworth: Penguin

ALTHUSSER, L. and BALIBAR, E. (1970), *Reading Capital*, London: New Left Books

AMSON, J. C. (1974), 'Equilibrium and catastrophic modes of urban growth', in E. L. Cripps (ed.), *Space–Time Concepts in Urban and Regional Models*, London: Pion, pp. 108–28

ANDERSON, J. (1973), 'Ideology in geography: an introduction', *Antipode*, vol. 5, no. 3, pp. 1–6

ANUCHIN, V. A. (1973), 'Theory of geography', in R. J. Chorley (ed.), *Directions in Geography*, London: Methuen

ARON, R. (1967), *Main Currents in Sociological Thought 2*, Harmondsworth: Penguin

BACHELARD, G. (1934), *Le Nouvel Esprit scientifique*, Paris: PUF

BADCOCK, C. (1975), *Lévi-Strauss: Structuralism and Sociological Theory*, London: Hutchinson

BAKER, A. R. H. (1976), 'On mental maps', *The Times Literary Supplement*, 17 December, p. 1582

BAKER, A. R. H. (1977), 'Rhetoric and reality in historical geography', *Journal of Historical Geography*, vol. 3, pp. 301–5

BARNBROCK, J. (1976), 'Prolegomenon to a methodological debate on location theory: the case of von Thünen', *Antipode*, vol. 6, no. 1, pp. 59–66

BARTELS, D. (1973), 'Between theory and metatheory', in R. J. Chorley (ed.), *Directions in Geography*, London: Methuen, pp. 25–42

BATTY, M. (1974), 'Spatial entropy', *Geographical Analysis*, vol. 6, pp. 1–31

BENNETT, R. J. (1974), 'Process identification for time series modelling in urban and regional planning', *Regional Studies*, vol. 8, pp. 157–74

BENNETT, R. J. (1975), 'Dynamic systems modelling of the North-west region', *Environment and Planning A*, vol. 7, pp. 525–38, 539–66, 617–36, 887–98

BENNETT, R. J. and CHORLEY, R. J. (1977), *Environmental Systems: Philosophy, Analysis, Control*, London: Methuen

BERGMANN, G. (1967), *The Metaphysics of Logical Positivism*, Madison: University of Wisconsin Press

BERNSTEIN, R. J. (1976), *The Restructuring of Social and Political Theory*, Oxford: Basil Blackwell

BERRY, B. J. L. (1970), 'The geography of the United States in the year 2000', *Transactions of the Institute of British Geographers*, vol. 51, pp. 21–53

BERRY, B. J. L. (1972), 'More on relevance and policy analysis', *Area*, vol. 4, pp. 77–80

BERRY, B. J. L. (1973), 'A paradigm for modern geography', in R. J. Chorley (ed.), *Directions in Geography*, London: Methuen

BHASKAR, R. (1975), 'Feyerabend and Bachelard: two philosophies of science', *New Left Review*, vol. 94, pp. 31–55

BILLINGE, M. (1977), 'In search of negativism: phenomenology and historical geography', *Journal of Historical Geography*, vol. 3, pp. 55–67

BIRD, J. (1975), 'Methodological implications for geography from the philosophy of K. R. Popper', *Scottish Geographical Magazine*, vol. 91, pp. 153–63

BIRD, J. (1977), 'Methodology and philosophy', *Progress in Human Geography*, vol. 1, pp. 104–10

BLACK, M. (1962), *Models and Metaphors*, Ithaca, N.Y.: Cornell University Press

BLAU, P. (1964), *Exchange and Power in Social Life*, New York: John Wiley

BLOWERS, A. (1972), 'Bleeding hearts and open values', *Area*, vol. 4, pp. 290–2

BOWEN, M. J. (1970), 'Mind and Nature: the physical geography of Alexander von Humboldt', *Scottish Geographical Magazine*, vol. 86, pp. 222–33

BROOKFIELD, H. C. (1975), *Interdependent Development*, London: Methuen

BROOKFIELD, H. C. and HART, D. (1971), *Melanesia: A Geographical Interpretation of an Island World*, London: Methuen

BRYANT, C. G. (1975), 'Positivism reconsidered', *Sociological Review*, vol. 23, pp. 397–412

BUNGE, W. (1962), *Theoretical Geography*, Lund: Gleerup

BUNGE, W. (1968), 'Fred K. Schaefer and the science of geography', *Harvard Papers in Theoretical Geography: Special Paper A*

BUNGE, W. (1973), 'The geography of human survival', *Annals of the Association of American Geographers*, vol. 63, pp. 275–95

BURGESS, R. (1976), 'Marxism and geography', *Occasional Papers, Department of Geography, University College London*, no. 30

BURTON, I. (1963), 'The quantitative revolution and theoretical geography', *Canadian Geographer*, vol. 7, pp. 151–62

BUTTIMER, A. (1971), *Society and Milieu in the French Geographic Tradition*, Chicago: Rand McNally

BUTTIMER, A. (1974), 'Values in geography', *Association of American Geographers, Commission on College Geography*, Resource Paper no. 24

BUTTIMER, A. (1976), 'Grasping the dynamism of the life-world', *Annals of the Association of American Geographers*, vol. 66, pp. 277–92

CAMPBELL, D. (1974), 'Role relations in advocacy geography', *Antipode*, vol. 6, no. 2, pp. 102–5

CARNAP, R. (1935), *Philosophy and Logical Syntax*, London: Kegan Paul

CARNEY, J., HUDSON, R., IVE, G. and LEWIS, J. (1976), 'Regional underdevelopment in late capitalism: a study of the North East of England', in I. Masser (ed.), *Theory and practice in regional science*', London: Pion, pp. 11–29

CASTELLS, M. (1972), *La question urbaine*, Paris: Maspero

CASTELLS, M. (1976), *La question urbaine* (2nd edition), Paris: Maspero

CASTELLS, M. (1977), *The Urban Question: a Marxist Approach*, London: Edward Arnold

CHAPMAN, G. P. (1970), 'The application of Information Theory to the analysis of population distributions in space', *Economic Geography*, vol. 46, pp. 317–31

CHAPMAN, G. P. (1977), *Human and Environmental Systems: A Geographer's Appraisal*, London: Academic Press

CHARBONNIER, G. (1969), *Conversations with Claude Lévi-Strauss*, London: Jonathan Cape

CHIARI, J. (1975), *Twentieth-Century French Thought: From Bergson to Lévi-Strauss*, London: Elek

CHISHOLM, M. (1967), 'General Systems Theory and geography', *Transactions of the Institute of British Geographers*, vol. 42, pp. 45–52

CHISHOLM, M. (1968), *Geography and Economics*, London: Bell

CHISHOLM, M. (1969), 'Social science research in geography', *Area*, vol. 1, pp. 8–9

CHISHOLM, M. (1971), 'In search of a basis for location theory: micro-economics or welfare economics', *Progress in Geography*, vol. 3, pp. 111–33

CHISHOLM, M. (1971a), 'Geography and the question of "relevance" ', *Area*, vol. 3, pp. 65–8

CHISHOLM, M. (1975), *Human Geography: Evolution or Revolution?*, Harmondsworth: Penguin

CHORLEY, R. J. (1971), 'The role and relations of physical geography', *Progress in Geography*, vol. 3, pp. 87–109

CHORLEY, R. J. (1973), 'Geography as human ecology', in R. J. Chorley (ed.), *Directions in Geography*, London: Methuen

CHORLEY, R. J. and HAGGETT, P. (eds) (1967), *Models in Geography*, London: Methuen

CHORLEY, R. J. and KATES, R. W. (1969), 'Introduction', to R. J. Chorley (ed.), *Water, Earth and Man: A Synthesis of Hydrology, Geomorphology and Socio-economic Geography*, London: Methuen

CHORLEY, R. J. and KENNEDY, B. A. (1971), *Physical Geography: A Systems Approach*, London: Prentice Hall

CHRISTALLER, W. (1933), *Die zentralen Orte in Süddeutschland*, Jena: Gustav Fischer; translated as *Central Places in Southern Germany*, Englewood Cliffs, N. J.: Prentice-Hall

CLARK, A. H. (1962), 'Praemia Geographiae: the incidental rewards of a geographical career', *Annals of the Association of American Geographers*, vol. 55, pp. 229–41

CLARK, W. and RUSHTON, G. (1970), 'Models of intra-urban consumer behavior and their implications for central place theory', *Economic Geography*, vol. 46, pp. 486–97

CLAVAL, P. (1975), 'Contemporary human geography in France', *Progress in Geography*, vol. 7, pp. 253–92

CLAVAL, P. and NARDY, J.-P. (1968), *Pour le cinquantenaire de la mort de Paul Vidal de la Blache: études d'histoire de la géographie*, Paris: Les Belles Lettres

CLIFF, A., HAGGETT, P., ORD, J. K., BASSETT, K. and DAVIES, R. L. (1975), *Elements of Spatial Structure: A Quantitative Approach*, Cambridge: Cambridge University Press

CLIFF, A. D. and ORD, J. K. (1975), 'Model building and the analysis of spatial pattern in human geography', *Journal of the Royal Statistical Society B*, vol. 37, pp. 297–348

CLIFF, A. D. and ORD, J. K. (1975), 'Space–time modelling with an application to regional forecasting', *Transactions of the Institute of British Geographers*, vol. 64, pp. 119–28

COLENUTT, R. (1976), 'Comment on "To what extent is the geographer's world the 'real' world?" ', *Area*, vol. 8, pp. 84–5

COLODNY, R. G. (ed.) (1970), *The Nature and Function of Scientific Theories*, Pittsburgh: Pittsburgh University Press

CONACHER, A. J. (1969), 'Open systems and dynamic equilibrium in geomorphology: a comment', *Australian Geographical Studies*, vol. 7, pp. 153–8

CONNELLY, D. S. (1972), 'Geomorphology and information theory, in R. J. Chorley (ed.), *Spatial analysis in geomorphology*, London: Methuen, pp. 91–108

CONNERTON, P. (ed.) (1976), *Critical Sociology*, Harmondsworth: Penguin

COOKE, R. U. and ROBSON, B. T. (1976), 'Geography in the United Kingdom, 1972–1976', *Geographical Journal*, vol. 142, pp. 81–100

COPPOCK, J. T. (1974), 'Geography and public policy: challenges, opportunities and implications', *Transactions of the Institute of British Geographers*, vol. 63, pp. 1–16

CULLEN, I. G. (1976), 'Human geography, regional science and the study of individual behaviour', *Environment and Planning A*, vol. 8, pp. 397–410

DARDEL, E. (1952), *L'Homme et la Terre: nature de réalité géographique*, Paris: PUF

DICKENSON, J. and CLARKE, C. G. (1972), 'Relevance and the "newest geography" ', *Area*, vol. 4, pp. 25–7

DICKINSON, R. E. (1969), *The Makers of Modern Geography*, London: Routledge & Kegan Paul

DOBB, M. (1973), *Theories of Value and Distribution since Adam Smith: Ideology and Economic Theory*, Cambridge: Cambridge University Press

DUHEM, P. (1954), *The Aim and Structure of Physical Theory*, Princeton: Princeton University Press

DUHEM, P. (1969), *To Save the Phenomena: An Essay on the Ideas of Physical Theory from Plato to Galileo*, Chicago: University of Chicago Press

DUNCAN, S. and SAYER, A. (1977), 'The "new" behavioural geography – a reply to Cullen', *Environment and Planning A*, vol. 9, pp. 230–2

DUPRÉ, G. and REY, P.-P. (1969), 'Réflexions sur la pertinence d'une théorie des échanges', *Cahiers Internationales de Sociologie*, vol. 46, pp. 133–62

DURKHEIM, E. (1893), *De la division du travail social: étude sur l'organisation des sociétés supérieures*, Paris: Félix Alcan; translated as *The Division of Labour in Society*, New York: Free Press

DURKHEIM, E. (1895), *Les Règles de la méthode sociologique*, Paris: Félix Alcan; translated as *The Rules of Sociological Method*, Chicago: University of Chicago Press

DURKHEIM, E. (1897–8), Review of *Politische Geographie*, *L'Année Sociologique*, vol. 2, pp. 522–32

DURKHEIM, E. (1898–9), Review of *Anthropogéographie*, *L'Année Sociologique*, vol. 3, pp. 550–8

DURKHEIM, E. (1912), *Les Formes élémentaires de la vie religieuse: le système totémique en Australie*, Paris: Félix Alcan; translated as *The Elementary Forms of the Religious Life*, London: George Allen & Unwin

EKEH, P. (1974), *Social Exchange Theory*, London: Heinemann

ENGELMANN, P. (1967), *Letters from Ludwig Wittgenstein*, Oxford: Oxford University Press

ENTRIKIN, J. N. (1977), 'Contemporary humanism in geography', *Annals of the Association of American Geographers*, vol. 66, pp. 615–32

FAY, B. (1975), *Social theory and political practice*, London: George Allen & Unwin

FEBVRE, L. (1932), *A Geographical Introduction to History*, London: Kegan Paul, Trench, Trübner

FEYERABEND, P. (1970), 'Against method: outline of an anarchistic theory of knowledge', *Minnesota Studies in the Philosophy of Science*, no. 4, Minneapolis: University of Minnesota Press

FEYERABEND, P. (1975), *Against Method*, London: New Left Books

FRANKEL, B. (1974), 'Habermas talking: an interview', *Theory and Society*, vol. 1, pp. 37–58

FRESHFIELD, D. (1886), 'The place of geography in education', *Proceedings of the Royal Geographical Society*, vol. 8, pp. 698–714

FREUND, J. (1968), *The Sociology of Max Weber*, Harmondsworth: Penguin

FRIEDMANN, J. (1972), 'The spatial organisation of power in the development of urban systems', *Comparative Urban Research*, vol. 1, pp. 5–42

FRIERE, P. (1972), *Pedagogy of the Oppressed*, Harmondsworth: Penguin

FRISBY, D. (1976), Introduction to T. Adorno, H. Albert, R. Dahrendorf, J. Habermas, H. Pilot and K. Popper, *The Positivist Dispute in German Sociology*, London: Heinemann

GADAMER, H.-G. (1975), *Truth and Method*, London: Sheed & Ward

GALE, S. (1972), 'Inexactness, Fuzzy sets and the foundations of behavioral geography', *Geographical Analysis*, vol. 4, pp. 337–49

GALE, S. (1972a), 'On the heterodoxy of explanation: a review of David Harvey's *Explanation in Geography*', *Geographical Analysis*, vol. 4, pp. 285–332

GALE, S. (1973), 'Explanation theory and models of migration', *Economic Geography*, vol. 49, pp. 257–74

GALOIS, B. (1976), 'Ideology and the idea of nature: the case of Peter Kropotkin', *Antipode*, vol. 8, no. 3, pp. 1–16

GARDNER, H. (1976), *The Quest for Mind: Piaget, Lévi-Strauss and the Structuralist Movement*, London: Quartet

GIDDENS, A. (1971), *Capitalism and Modern Social Theory*, Cambridge: Cambridge University Press

GIDDENS, A. (ed.) (1972), *Emile Durkheim: Selected Writings*, Cambridge: Cambridge University Press

GIDDENS, A. (1976),   *New Rules of Sociological Method: A Positive Critique of Interpretative Sociologies*, London: Hutchinson

GILBERT, E. W. and STEEL, R. W. (1945), 'Social geography and its place in colonial studies', *Geographical Journal*, vol. 106, pp. 118–31

GIRT, J. L. (1976), 'Some extensions to Rushton's spatial preference scaling model', *Geographical Analysis*, vol. 8, pp. 137–52

GLACKEN, C. (1967),   *Traces on the Rhodian Shore: Nature and Culture in Western Thought from Ancient Times to the End of the Eighteenth Century*, Berkeley: University of California Press

GLUCKSMANN, A. (1972), 'A ventriloquist structuralism' in New Left Review (ed.), *Western Marxism: a critical reader*, London: New Left Books, pp. 282–314

GLUCKSMANN, M. (1974),   *Structuralist Analysis in Contemporary Social Thought*, London: Routledge & Kegan Paul

GODDARD, D. (1975), 'Philosophy and structuralism', *Philosophy of the Social Sciences*, vol. 5, pp. 103–23

GODELIER, M. (1973),   *Horizon, trajets marxistes en anthropologie*, Paris: Maspero

GOLLEDGE, R. and AMADEO, D. (1968), 'On laws in geography', *Annals of the Association of American Geographers*, vol. 58, pp. 760–74

GORMAN, R. A. (1977),   *The Dual Vision: Alfred Schutz and the Myth of a Phenomenological Social Science*, London: Routledge & Kegan Paul

GOULD, P. (1974), 'Some Steineresque comments and Monodian asides on geography in Europe', *Geoforum*, vol. 17, pp. 9–13

GOULDNER, A. (1971),   *The Coming Crisis of Western Sociology*, London: Heinemann

GOULDNER, A. (1976),   *The Dialectic of Ideology and Technology: The Origins, Grammar and Future of Ideology*, London: Macmillan

GREGORY, D. (1976), 'Rethinking historical geography', *Area*, vol. 8, pp. 295–9

GREGORY, D. (1977), 'Alfred Weber and location theory: a re-examination', paper presented at a colloquium organised by the Commission on the History of Geographical Thought of the International Geographical Union, University of Edinburgh

GREGORY, D. (1978), 'The discourse of the past: phenomenology, structuralism and historical geography', *Journal of Historical Geography*, vol. 4, in press

GREGORY, D. and SMITH, R. M. (1977), 'Polarities and lacunae: town–country relations in the transition from feudalism to capitalism', mimeo

GUELKE, L. (1971), 'Problems of scientific explanation in geography', *Canadian Geographer*, vol. 15, pp. 38–53

GUELKE, L. (1974), 'An idealist alternative in human geography', *Annals of the Association of American Geographers*, vol. 64, pp. 193–202

HABERMAS, J. (1972), *Knowledge and Human Interests*, London: Heinemann

HABERMAS, J. (1974), *Theory and Practice*, London: Heinemann

HABERMAS, J. (1976), 'The analytical theory of science and dialectics', in T. Adorno, H. Albert, R. Dahrendorf, J. Habermas, H. Pilot and K. Popper, *The Positivist Dispute in German Sociology*, London: Heinemann

HABERMAS, J. (1976a), *Legitimation Crisis*, London: Heinemann

HÄGERSTRAND, T. (1973), 'The domain of human geography', in R. J. Chorley (ed.), *Directions in Geography*, London: Methuen

HAGGETT, P. and CHORLEY, R. J. (1969), *Network Analysis in Geography*, London: Edward Arnold

HAGGETT, P., CLIFF, A. D. and FREY, A. (1977), *Locational Analysis in Human Geography*, London: Edward Arnold

HARRE, R. (1972), *The Philosophies of Science: An Introductory Survey*, London: Oxford University Press

HARRIS, C. (1971), 'Theory and synthesis in human geography', *Canadian Geographer*, vol. 15, pp. 157–72

HARRIS, C. (1975), Review of *Topophilia, Canadian Geographer*, vol. 19, pp. 163–4

HARTSHORNE, R. (1939), *The Nature of Geography: A Critical Survey of Current Thought in the Light of the Past*, Lancaster, Penn.: Association of American Geographers

HARTSHORNE, R. (1955), ' "Exceptionalism in geography" reexamined', *Annals of the Association of American Geographers*, vol. 45, pp. 205–44

HARVEY, D. (1969), *Explanation in Geography*, London: Edward Arnold

HARVEY, D. (1972), 'On obfuscation in geography: a comment on Gale's heterodoxy', *Geographical Analysis*, vol. 4, pp. 323–30

HARVEY, D. (1973), *Social Justice and the City*, London: Edward Arnold

184   *Bibliography*

HARVEY, D. (1974), Review of *The Human Consequences of Urbanism, Annals of the Association of American Geographers*, vol. 65, pp. 99–103

HARVEY, D. (1974a), 'What kind of geography for what kind of public policy?', *Transactions of the Institute of British Geographers*, vol. 63, pp. 18–24

HARVEY, D. (1975), 'Class structure in a capitalist society and the theory of residential differentiation', in R. Peel, P. Haggett and M. Chisholm (eds), *Processes in Physical and Human Geography: Bristol Essays*, London: Heinemann

HAWTHORN, G. (1976), *Enlightenment and Despair: A History of Sociology*, Cambridge: Cambridge University Press

HEMPEL, C. G. (1952), 'Fundamentals of concept formation in empirical science', *International Encyclopaedia of Unified Science*, vol. 2, no. 7

HEMPEL, C. G. (1965), *Aspects of Scientific Explanation*, New York: The Free Press

HESSE, M. (1974), *The Structure of Scientific Inference*, London: Macmillan

HINDESS, B. (1977), *Philosophy and Methodology in the Social Sciences*, Hassocks: Harvester Press

HINDESS, B. and HIRST, P. (1975), *Pre-Capitalist Modes of Production*, London: Routledge & Kegan Paul

HINDESS, B. and HIRST, P. (1977), *Mode of Production and Social Formation: Auto-Critique of Pre-Capitalist Modes of Production*, London: Macmillan

HIRST, P. (1975), *Durkheim, Bernard and Epistemology*, London: Routledge & Kegan Paul

HOBSBAWM, E. (1972), 'Karl Marx's contribution to historiography', in R. Blackburn (ed.), *Ideology in Social Science: Readings in Critical Social Theory*, London: Fontana, pp. 265–8:

HODGSHON, G. (1976), 'Exploitation and embodied labour time' *Bulletin of the Conference of Socialist Economists*, vol. 5, no. 1. pp. GH 1–23

HOLLIS, M. and NELL, E. (1975), *Rational Economic Man: A Philosophical Critique of Neoclassical Economics*, Cambridge: Cambridge University Press

HOOVER, E. M. (1948), *The Location of Economic Activity*, New York: McGraw Hill

HORKHEIMER, M. (1972), *Critical Theory*, New York: Seabury Press

Bibliography 185

HORVARTH, R. J. (1971), 'The Detroit Geographical Expedition and Institute experience', *Antipode*, vol. 3, no. 1, pp. 73–85

HOWARD, M. C. and KING, J. E. (1975), *The Political Economy of Marx*, London: Longmans

HUSSERL, E. (1970), *The Crisis of European Sciences and Transcendental Phenomenology*, Evanston: Northwestern University Press

ISARD, W. (1956), *Location and Space-Economy*, Cambridge, Mass.: MIT Press

ISARD, W. and LIOSSATOS, P. (1975), 'Parallels from physics for space–time development models, part 1', *Regional Science and Urban Economics*, vol 5, pp. 5–40

ISARD, W. and LIOSSATOS, P. (1975a), 'Parallels from physics for space–time development models, part II: interpretations and extensions of the basic model', *Papers of the Regional Science Association*, vol. 34, pp. 43–66

IVE, G. (1975), 'Walker and the "new conceptual framework" of urban rent', *Antipode*, vol. 7, no. 1, pp. 20–30

JAY, M. (1973), *The Dialectical Imagination*, London: Heinemann

JENNINGS, J. N. (1973), ' "Any millenniums today, lady?" The geomorphic bandwaggon parade', *Australian Geographical Studies*, vol. 11, pp. 115–33

KASPERSON, R. E. and BREITBART, M. (1974), 'Participation, decentralization and advocacy planning', *Association of American Geographers, Commission on College Geography*, Resource Paper no. 25

KEAT, R. and URRY, J. (1975), *Social Theory as Science*, London: Routledge & Kegan Paul

KELTIE, J. S. (1897), 'Some geographical problems', *Geographical Journal*, vol. 10, pp. 308–23

KING, L. J. (1976), 'Alternatives to a positive economic geography', *Annals of the Association of American Geographers*, vol. 66, pp. 293–308

KIRK, W. (1951), 'Historical geography and the concept of the Behavioural Environment', *Indian Geographical Journal*, Silver Jubilee volume, pp. 152–60

186 Bibliography

KIRK, W. (1963), 'Problems of geography', *Geography*, vol. 48, pp. 357–71

KOCKELMANS, J. J. (1966), *Phenomenology and Physical Science*, Pittsburgh: Duquesne University Press

KOLAKOWSKI, L. (1972), *Positivist Philosophy: From Hume to the Vienna Circle*, Harmondsworth: Penguin

KRAFT, V. (1953), *The Vienna Circle: The Origin of Neo-Positivism*, New York: Philosophical Library

KUHN, T. S. (1962), *The Structure of Scientific Revolutions*, Chicago: University of Chicago Press

KUHN, T. S. (1970), *The Structure of Scientific Revolutions* (2nd edition), Chicago: University of Chicago Press

LACLAU, E. (1975), 'The specificity of the political: the Poulantzas–Miliband debate', *Economy and Society*, vol. 4, pp. 87–110

LAKATOS, I. (1970), 'Falsification and the methodology of scientific research programmes', in I. Lakatos and H. Musgrave (eds), *Criticism and the Growth of Knowledge*, Cambridge: Cambridge University Press

LANGTON, J. (1972), 'Potentialities and problems of adopting a systems approach to the study of change in human geography', *Progress in Geography*, vol. 4, pp. 125–79

LEACH, B. (1974), 'Race, problems and geography', *Transactions of the Institute of British Geographers*, vol. 63, pp. 41–7

LECOURT, D. (1975), *Marxism and Epistemology: Bachelard, Canguilhem, Foucault*, London: New Left Books

LEFEBVRE, H. (1974), *La production de l'espace*, Paris: Anthropos

LENHARDT, C. (1976), 'The wanderings of enlightenment', in J. O'Neill (ed.), *On Critical Theory*, London: Heinemann

LEOPOLD, L. and LANGBEIN, W. B. (1962), 'The concept of entropy in landscape evolution', *United States Geological Survey, Professional Paper* 500A

LÉVI-STRAUSS, C. (1955), *Tristes Tropiques*, Paris: Plon

LÉVI-STRAUSS, C. (1960), 'On manipulated sociological models', *Bijdragen tot de taal-, land- en volkenkunde*, vol. 116, pp. 45–54

LÉVI-STRAUSS, C. (1963), 'The bear and the barber', *Journal of the Royal Anthropological Institute*, vol. 93, pp. 1–11

LÉVI-STRAUSS, C. (1964), *Totemism*, London: Merlin Press

LÉVI-STRAUSS, C. (1964), *Mythologiques I: le cru et le cuit*, Paris: Plon

LÉVI-STRAUSS, C. (1966), *The Savage Mind*, London: Weidenfeld & Nicolson

LÉVI-STRAUSS, C. (1967), *The Scope of Anthropology*, London: Jonathan Cape

LÉVI-STRAUSS, C. (1969), *The Elementary Structures of Kinship*, London: Eyre & Spottiswoode

LÉVI-STRAUSS, C. (1973), *From Honey to Ashes*, London: Jonathan Cape

LÉVI-STRAUSS, C. (1977), *Structural Anthropology 2*, London: Allen Lane

LEWIS, J. and MELVILLE, B. (in press), 'The politics of epistemology in regional science', in P. W. J. Batey (ed.), *Theory and Method in Urban and Regional Analysis*, London: Pion, in press

LEY, D. (1974), 'The city and good and evil: reflections on Christian and Marxian interpretations', *Antipode*, vol. 6, no. 1, pp. 66–74

LIPIETZ, A. (1975), 'Structuration de l'espace problème foncier et aménagement du territoire', *Environment and Planning A*, vol. 7, pp. 415–25

LOBKOWICZ, N. (1972), 'Interest and objectivity', *Philosophy of the Social Sciences*, vol. 2, pp. 193–210

LÖSCH, A. (1954), *The Economics of Location*, Yale: Yale University Press

LOWENTHAL, D. (1961), 'Geography, experience and imagination: towards a geographical epistemology', *Annals of the Association of American Geographers*, vol. 51, pp. 241–60

LOWENTHAL, D. and PRINCE, H. C. (1976), 'Transcendental experience', in S. Wapner, S. Cohen and B. Kaplan (eds), *Experiencing the Environment*, New York: Plenum

LUKERMANN, F. (1964), 'Geography as a formal intellectual discipline and the way in which it contributes to human knowledge', *Canadian Geographer*, vol. 8, pp. 167–72

LUKES, S. (1973), *Emile Durkheim: His Life and Work*, London: Allen Lane

MACINTYRE, S. and TRIBE, K. (1975), *Althusser and Marxist Theory*, Cambridge: Macintyre & Tribe

MACKINDER, H. J. (1887), 'On the scope and methods of geography', *Proceedings of the Royal Geographical Society*, vol. 9, pp. 141–60

MALINOWSKI, B. (1922), *Argonauts of the Western Pacific*, London: Routledge & Kegan Paul

MANN, H.-D. (1971), *Lucien Febvre: la pensée vivante d'un historien*, Paris: Colin

MANUEL, F. (1962), *The Prophets of Paris*, Cambridge, Mass.: Harvard University Press

MARCHAND, B. (1972), 'Information theory and geography', *Geographical Analysis*, vol. 4, pp. 234–57

MARCUSE, H. (1972), *One Dimensional Man*, London: Abacus

MARKHAM, C. R. (1898), 'The field of geography', *Geographical Journal*, vol. 11, pp. 1–15

MARTIN, R. L. (1977), 'Review of *The New Urban Economics and Alternatives*', *Environment and Planning A*, vol. 9, pp. 1081–3

MARTIN, R. L. and OEPPEN, J. E. (1975), 'The identification of regional forecasting models using space–time correlation functions', *Transactions of the Institute of British Geographers*, vol. 66, pp. 95–118

MARX, K. (1976), *Capital: A Critique of Political Economy, 1*, Harmondsworth:Penguin

MASSEY, D. (1973), 'Towards a critique of industrial location theory', *Antipode*, vol. 5, no. 3, pp. 33–9

MAUSS, M. (1970), *The Gift: Forms and Functions of Exchange in Archaic Societies*, London: Routledge & Kegan Paul

MCCARTHY, T. A. (1973), 'A theory of communicative competence', *Philosophy of the Social Sciences*, vol. 3, pp. 135–56

MEDVEDKOV, Y. (1970), 'Entropy: an assessment of its potentialities in geography', *Economic Geography*, vol. 46, pp. 306–16

MEES, A. I. (1975), 'The revival of cities in medieval Europe: an application of catastrophe theory', *Regional Science and Urban Economics*, vol 5, pp. 403–25

MEPHAM, J. (1973), 'The structuralist sciences and philosophy', in D. Robey (ed.), *Structuralism: An Introduction*, Oxford: Oxford University Press

MILLS, E. (1972), *Studies in the Structure of the Urban Economy*, Baltimore: Johns Hopkins University Press

MOORE, E. G. and GALE, S. (1973), 'Comments on models of occupancy patterns and neighborhood change', in E. G. Moore (ed.), *Models of Residential Location and Relocation in the City, Northwestern University Studies in Geography*, no. 20

MUTH, R. (1969), *Cities and Housing*, Chicago: Chicago University Press

NIJKAMP, P. and PAELINCK, J. (1973), 'A solution method for

neoclassical location problems', *Regional and Urban Economics*, vol. 3, pp. 383–410

NORDBECK, S. (1965), 'The law of allometric growth', *Michigan Inter-University Community of Mathematical Geographers, Discussion Paper*, no. 7

NORMAN, R. (1976), 'On dialectic', *Radical Philosophy*, vol. 14, pp. 2–9

OLLMAN, B. (1971), *Alienation: Marx's Conception of Man in Capitalist Society*, Cambridge: Cambridge University Press

OLSSON, G. (1971), 'Correspondence rules and social engineering', *Economic Geography*, vol. 47, pp. 545–54

OLSSON, G. (1972), 'Some notes on geography and social engineering', *Antipode*, vol. 4, no. 1, pp. 1–22

OLSSON, G. (1974), 'The dialectics of spatial analysis', *Antipode*, vol. 6, no. 3, pp. 50–62

OLSSON, G. (1975), *Birds in Egg*, Michigan Geographical Publications, no. 15

OVERTON, D. (1976), 'Towards a discussion of the nature of radical geography', *Antipode*, vol. 8, no. 3, pp. 86–7

VAN PAASSEN, C. (1976), 'Human geography in terms of existential anthropology', *Tijdschrift voor Economische en Sociale Geografie*, vol. 67, pp. 324–41

PAHL, R. E. (1967), 'Sociological models in geography', in R. J. Chorley and P. Haggett (eds), *Models in Geography*, London: Methuen

PAHL, R. E. (1975), *Whose City? And Further Essays on Urban Society*, Harmondsworth: Penguin

PALANDER, T. (1935), *Beiträge zur Standortstheorie*, Uppsala: Almqvist & Wiksells

PALMER, R. (1969), *Hermeneutics*, Evanston: Northwestern University Press

PEET, R. (1975), 'Inequality and poverty: a Marxist-geographic theory', *Annals of the Association of American Geographers*, vol. 65, pp. 564–71

PIAGET, J. (1971), *Structuralism*, London: Routledge & Kegan Paul

PICKVANCE, C. G. (ed.), *Urban Sociology: Critical Essays*, London: Tavistock

PINKARD, T. (1976), 'Interpretation and verification in the human sciences: a note on Taylor', *Philosophy of the Social Sciences*, vol. 6, pp. 165–73

PIVČEVIĆ, E. (1970), *Husserl and Phenomenology*, London: Hutchinson

POLANYI, K. (1968), *Primitive, Archaic and Modern Economies: Essays of Karl Polanyi*, Boston: Beacon Press

POPPER, K. R. (1959), *The Logic of Scientific Discovery*, London: Hutchinson

POPPER, K. R. (1976), *Unended Quest: An Intellectual Autobiography*, London: Fontana

POPPER, K. (1976a), 'The logic of the social sciences', in T. Adorno, H. Albert, R. Dahrendorf, J. Habermas, H. Pilot, K. Popper, *The Positivist Dispute in German Sociology*, London: Heinemann

POSTER, M. (1976), *Existential Marxism in Post-War France: From Sartre to Althusser*, Princeton: Princeton University Press

POULANTZAS, N. (1973), *Political Power and Social Classes*, London: New Left Books and Sheed & Ward

PRED, A. (1973), 'Urbanisation, domestic planning problems and Swedish geographic research', *Progress in Geography*, vol. 5, pp. 1–76

PRED, A. (1977), 'The choreography of existence: comments on Hägerstrand's time-geography', *Economic Geography*, vol. 53, pp. 207–21

PRINCE, H. C. (1971), 'Questions of social relevance', *Area*, vol. 3, pp. 150–3

QUINE, W. V. O. (1961), *From a Logical Point of View*, Cambridge, Mass.: Harvard University Press

RATZEL, F. (1882), *Anthropogeographie oder Grundzüge der Anwendung der Erdkunde auf die Geschichte*, Stuttgart: Engelhorn

RATZEL, F. (1891), *Anthropogeographie, 2 Teil: Die geographische Verbreitung der Menschen*, Stuttgart: Engelhorn

RATZEL, F. (1897), *Politische Geographie*, Münich and Leipzig: Oldenburg

RATZEL, F. (1898–9), 'Le sol, la société et l'état', *L'Année Sociologique*, vol. 3, pp. 1–4

REIN, M. (1976), *Social Science and Public Policy*, Harmondsworth: Penguin

RELPH, E. (1970), 'An inquiry into the relations between phenomenology and geography', *Canadian Geographer*, vol. 14, pp. 193–201

RELPH, E. (1976), *Place and Placelessness*, London: Pion

RICOUER, P. (1965), *Freud and Philosophy: An Essay on Interpretation*, New Haven: Yale University Press

ROBERTSON, G. S. (1900), 'Political geography and the Empire', *Geographical Journal*, vol. 16, pp. 447–57

ROBSON, B. T. (1971), 'Down to earth', *Area*, vol. 3, p. 137

ROBSON, B. T. (1972), 'The corridors of geography', *Area*, vol. 4, pp. 213–14

ROWTHORN, R. E. (1974), 'Neo-classicism, neo-Ricardianism and Marxism', *New Left Review*, vol. 86, pp. 63–87

RUSHTON, G. (1969), 'Analysis of spatial behavior by revealed space preference', *Annals of the Association of American Geographers*, vol. 59, pp. 391–400

SACK, D. (1972), 'Geography, geometry and explanation', *Annals of the Association of American Geographers*, vol. 62, pp. 61–78

SACK, D. (1974), 'Chorology and spatial analysis', *Annals of the Association of American Geographers*, vol. 64, pp. 439–52

SAHLINS, M. (1974), *Stone Age Economics*, London: Tavistock

SANTOS, M. (1977), 'Society and space: social formation as theory and method', *Antipode*, vol. 9, no. 1, pp. 3–13

SANTOS, M. and PEET, R. (1977), 'Introduction: underdevelopment in the Third World: socio-economic formation and spatial organisation', *Antipode*, vol 9, no. 1, pp. 1–3

SAUSSURE, F. (1916), *Cours de linguistique générale*, Paris: Payot

SAYER, R. A. (1976), 'A critique of urban modelling: from regional science to urban and regional political economy', *Progress in Planning*, vol. 6, no. 3

SCHAEFER, F. K. (1953), 'Exceptionalism in geography: a methodological examination', *Annals of the Association of American Geographers*, vol. 43, pp. 226–49

SCHEFFLER, I. (1967), *Science and Subjectivity*, Indianapolis: Bobbs-Merrill

SCHUTZ, A. (1962), *Collected Papers*, The Hague: Martinus Nijhoff

SCHUTZ, A. (1967), *The Phenomenology of the Social World*, Evanston: Northwestern University Press

SCOTT, A. J. (1976), 'Land and land rent: an interpretative review of the French literature', *Progress in Geography*, vol. 9, pp. 101–45

SCOTT, A. J. (1976a), 'Review of *Birds in Egg*', *Annals of the Association of American Geographers*, vol. 66, pp. 633–6

SHORT, J. (1974), 'Social systems and spatial patterns', *Antipode*, vol. 8, no. 1, pp. 77–83

SMART, B. (1976), *Sociology, Phenomenology and Marxian Analysis: A Critical Discussion of the Theory and Practice of a Science of Society*, London: Routledge & Kegan Paul

SMITH, D. M. (1971), 'Radical geography: the next revolution?', *Area*, vol. 3, pp. 153–7

SMITH, D. M. (1973), 'Alternative "relevant" professional roles', *Area*, vol. 5, pp. 1–4

SMITH, D. M. (1977), *Human Geography: A Welfare Approach*, London: Edward Arnold

STAMP, L. D. (1951), 'Applied geography', in L. D. Stamp and S. W. Wooldridge (eds), *London Essays in Geography*, London: Longmans

STEDMAN JONES, G. (1974), 'Working-class culture and working-class politics in London, 1870–1900: notes on the remaking of a working class', *Journal of Social History*, vol. 7, pp. 460–508

STEDMAN JONES, G. (1977), 'Comment on "Labour, capital and class struggle around the built environment"', paper presented at a symposium organized by the Historical Geography Research Group, University of Newcastle-upon-Tyne

STEINER, G. (1975), *After Babel: Aspects of Language and Translation*, London and New York: Oxford University Press

STODDART, D. R. (1967), 'Organism and ecosystem as geographical models', in R. J. Chorley and P. Haggett (eds), *Models in Geography*, London: Methuen

STODDART, D. R. (1975), 'Kropotkin, Reclus and "relevant" geography', *Area*, vol. 7, pp. 188–190

STODDART, D. R. (1975a), 'The RGS and the foundations of geography at Cambridge', *Geographical Journal*, vol. 141, pp. 216–39

STODDART, D. R. (1977), 'The paradigm concept and the history of geography', paper presented at a colloquium organized by the Commission on the History of Geographical Thought of the International Geographical Union, University of Edinburgh

STRACHEY, R. (1888), 'Lectures on geography delivered before the University of Cambridge, I', *Proceedings of the Royal Geographical Society*, vol. 10, pp. 146–60

TAYLOR, C. (1971), 'Interpretation and the science of man', *Review of Metaphysics*, vol. 25, pp. 3–51

THRIFT, N. (1977), 'Time and theory in human geography', *Progress in Human Geography*, vol. 1, pp. 65–101

THRIFT, N. (1977a), *An Introduction to Time Geography*, Norwich: Geo Abstracts

TUAN, YI-FU (1971), 'Geography, phenomenology and the study of human nature', *Canadian Geographer*, vol. 15, pp. 181–92

TUAN, YI-FU (1972), 'Structuralism, existentialism and environmental perception', *Environment and Behavior*, vol. 3, pp. 319–31

TUAN, YI-FU (1974), 'Space and place: humanistic perspective', *Progress in Geography*, vol. 6, pp. 212–52

TUAN, YI-FU (1974a), *Topophilia: A Study of Environmental Perception, Attitudes and Values*, Englewood Cliffs: Prentice-Hall

TUAN, YI-FU (1975), 'Images and mental maps', *Annals of the Association of American Geographers*, vol. 65, pp. 205–13

TUAN, YI-FU (1976), 'Humanistic geography', *Annals of the Association of American Geographers*, vol. 66, pp. 266–76

VIDAL DE LA BLACHE, P. (1911), 'Les genres de vie dans la géographie humaine', *Annales de Géographie*, vol. 20, pp. 193–212

VIDAL DE LA BLACHE, P. (1922), *Principes de géographie humaine*, Paris: Colin

VILAR, P. (1973), 'Histoire marxiste, histoire en construction: essai de dialogue avec Althusser', *Annales ESC*, vol. 28, pp. 165–98

VON BERTALANFFY, L. (1973), *General System Theory: Foundations, Development, Applications*, Harmondsworth: Penguin

WALLWORK, E. (1972), *Durkheim: Morality and Milieu*, Cambridge, Mass.: Harvard University Press

WARNTZ, W. (1973), 'New geography as general spatial systems theory – old social physics writ large?', in R. J. Chorley (ed.), *Directions in Geography*, London: Methuen

WEBER, A. (1909), *Über den Standort der Industrien, I: Reine Theorie des Standorts*, Tübingen; translated as *Alfred Weber's Theory of the Location of Industries*, New York: Russell & Russell

WELLMER, A. (1971), *Critical Theory of Society*, New York: Herder & Herder

WELLMER, A. (1976), 'Communications and emancipation: reflections on the linguistic turn in critical theory', in J. O'Neill (ed.), *On Critical Theory*, London: Heinemann

WHITE, G. (1972), 'Geography and public policy', *Professional Geographer*, vol. 24, pp. 101–4

WILSON, A. G. (1970), *Entropy in Urban and Regional Modelling*, London: Pion

WILSON, A. G. (1972), 'Theoretical geography: some speculations', *Transactions of the Institute of British Geographers*, vol. 57, pp. 31–44

WILSON, A. G. (1974), *Urban and Regional Models in Geography and Planning*, London: Wiley

WILSON, A. G. (1976), 'Catastrophe theory and urban modelling: an application to modal choice', *Working Papers, Department of Geography, University of Leeds*, no. 133

WISE, M. J. (1977), 'On progress in geography', *Progress in Human Geography*, vol. 1, pp. 1–11

WOLDENBERG, M. (1970), 'The hexagon as a spatial average', *Harvard Papers in Theoretical Geography*, no. 42

WOLDENBERG, M. (1972), 'The average hexagon in spatial hierarchies', in R. J. Chorley (ed.), *Spatial Analysis in Geomorphology*, London: Methuen

WOLDENBERG, M. and BERRY, B. J. L. (1967), 'Rivers and central places: analogous systems?', *Journal of Regional Science*, vol. 7, pp. 129–39

WOLFF, J. (1975), *Hermeneutic Philosophy and the Sociology of Art*, London: Routledge & Kegan Paul

WRIGHT, J. K. (1966), *Human Nature in Geography: Fourteen Papers, 1925–65*, Cambridge, Mass.: Harvard University Press

ZELINSKY, W. (1975), 'The demigod's dilemma', *Annals of the Association of American Geographers*, vol. 65, pp. 123–43

Index